EVOLUTION
AND
ETHICS

SCIENCE
AND
MORALS

Titles on Science in
Prometheus's Great Minds Series

See the back of this volume for a complete list of titles in Prometheus's
Great Books in Philosophy and Great Minds series.

EVOLUTION
AND
ETHICS

SCIENCE
AND
MORALS

THOMAS H.
HUXLEY

GREAT MINDS SERIES

 **Prometheus Books**

59 John Glenn Drive
Amherst, New York 14228-2197

Published 2004 by Prometheus Books

59 John Glenn Drive, Amherst, New York 14228–2197,
716–691–0133. FAX: 716–691–0137.

Library of Congress Cataloging-in-Publication Data

Huxley, Thomas Henry, 1825–1895.
Evolution and ethics ; and, Science and morals / by Thomas
H. Huxley.
     p. cm. — (Great minds series)
Originally published: New York : D. Appleton, 1896.
ISBN 1–59102–126–X
    1. Ethics, Evolutionary. I. Title: Evolution and ethics ; and,
Science and morals. II. Huxley, Thomas Henry, 1825–1895.
Science and morals. III. Title: Science and morals. IV. Title.
V. Series.

BJ1311.H8 2003
171'.7—dc22   2003068911

Printed in the United States of America on acid-free paper.

THOMAS HENRY HUXLEY was born on May 4, 1825, in Ealing, England. Although the son of a schoolmaster, Huxley had no formal education as a child; he read voraciously, however, and at a young age began to study medicine. Later, he entered Charing Cross Hospital medical school, taking his degree in 1845.

After passing the Royal College of Surgeons examination in 1846, Huxley was appointed assistant surgeon aboard the HMS *Rattlesnake* on its four-year scientific exploration of the southern seas around Australia. During that time, Huxley made extensive studies of the local marine life, which were later published to great acclaim. These marine studies, as well as Huxley's detailed investigations into comparative anatomy, paleontology, and evolution, confirmed forever his reputation as one of England's foremost scientists and controversial figures.

Huxley met Charles Darwin in 1851; after the publication of *The Origin of Species* in 1859, Huxley became Darwin's principal defender against the anti-evolutionists, including the Duke of Argyll, William Gladstone, and Bishop Samuel Wilberforce, whom Huxley engaged in a fierce debate over evolution in 1860. In 1863, Huxley published his *Evidence as to Man's Place in Nature*, which argued that human beings were closely related to anthropoid apes. Unorthodox both in science and religion, Huxley attacked what he called "that clericalism, which in England, as everywhere else, and to whatever denomination it may belong, is the deadly enemy of science." In a series of spirited essays and lectures on philosophy, religion, and science delivered in England and abroad, Huxley denounced orthodoxy and biblical infallibility; in light of the fact that no irrefutable evidence of the unseen world of religion could be adduced, Huxley espoused a healthy agnosticism concerning the supernatural.

For his work in science, Huxley held several academic positions concurrently and reaped many honors: from 1854 until 1885, he was Lecturer at the Royal School of Mines; Hunterian Professor at the Royal College of Surgeons (1863–1869); and Fullerian Professor at the Royal Institution (1863–1867). In 1883, Huxley was elected president of the Royal Society, a post he held for two years. A tireless popularizer of science as well as a specialist, Huxley served from 1870 to 1872 on the first London school board, and did much to promote educational techniques and the study of biology. Thomas Henry Huxley died on June 29, 1895, in Eastbourne, England.

Huxley's other major works include *Introduction to the Classification of Animals* (1869), *Lay Sermons* (1870), *Manual of the Comparative Anatomy of Vertebrated Animals* (1871), *Science and Morals* (1886), and *Evolution and Ethics* (1893).

PREFACE

THE discourse on "Evolution and Ethics," reprinted in the first half of the present volume, was delivered before the University of Oxford, as the second of the annual lectures founded by Mr. Romanes: whose name I may not write without deploring the untimely death, in the flower of his age, of a friend endeared to me, as to so many others, by his kindly nature; and justly valued by all his colleagues for his powers of investigation and his zeal for the advancement of knowledge. I well remember, when Mr. Romanes' early work came into my hands, as one of the secretaries of the Royal Society, how much I rejoiced in the accession to the ranks of the little army of workers in science of a recruit so well qualified to take a high place among us.

It was at my friend's urgent request that I agreed to undertake the lecture, should I be honoured with an official proposal to give it, though I confess not without misgivings, if only on account of the serious fatigue and hoarseness which public speaking has for some years caused me; while I knew that it would be my fate to follow the most accomplished and facile orator of our time, whose indomitable youth is in no matter more manifest than in his penetrating and musical voice. A certain saying about comparisons intruded itself somewhat importunately.

And even if I disregarded the weakness of my body in the matter of voice, and that of my mind in the matter of vanity, there remained a third difficulty. For several reasons, my attention, during a number of years, has been much directed to the bearing of modern scientific thought on the problems of morals and of politics, and I did not care to be diverted from that topic. Moreover, I thought it the most important and the worthiest which, at the present time, could engage the attention even of an ancient and renowned University.

But it is a condition of the Romanes foundation that the lecturer shall abstain from treating of either Religion or Politics; and it appeared to me that, more than most, perhaps, I was bound to act, not merely up to the letter, but in the spirit, of that prohibition. Yet Ethical Science is, on all sides, so entangled with Religion and Politics, that the lecturer who essays to touch the former without coming into contact with either of the latter, needs all the dexterity of an egg-dancer; and may even discover that his sense of clearness

and his sense of propriety come into conflict, by
no means to the advantage of the former.

I had little notion of the real magnitude of
these difficulties when I set about my task; but I
am consoled for my pains and anxiety by observing
that none of the multitudinous criticisms with
which I have been favoured and, often, instructed,
find fault with me on the score of having strayed
out of bounds.

Among my critics there are not a few to whom
I feel deeply indebted for the careful attention
which they have given to the exposition thus
hampered; and further weakened, I am afraid, by
my forgetfulness of a maxim touching lectures of
a popular character, which has descended to me
from that prince of lecturers, Mr. Faraday. He
was once asked by a beginner, called upon to
address a highly select and cultivated audience,
what he might suppose his hearers to know
already. Whereupon the past master of the art of
exposition emphatically replied " Nothing ! "

To my shame as a retired veteran, who has all
his life profited by this great precept of lec-
turing strategy, I forgot all about it just when
it would have been most useful. I was fatuous
enough to imagine that a number of propositions,
which I thought established, and which, in fact, I
had advanced without challenge on former oc-
casions, needed no repetition.

I have endeavoured to repair my error by

prefacing the lecture with some matter—chiefly
elementary or recapitulatory—to which I have
given the title of "Prolegomena." I wish I could
have hit upon a heading of less pedantic aspect
which would have served my purpose; and if it
be urged that the new building looks over large
for the edifice to which it is added, I can only
plead the precedent of the ancient architects,
who always made the adytum the smallest part
of the temple.

If I had attempted to reply in full to the
criticisms to which I have referred, I know not
what extent of ground would have been covered
by my *pronaos*. All I have endeavoured to do,
at present, is to remove that which seems to
have proved a stumbling-block to many—namely,
the apparent paradox that ethical nature, while
born of cosmic nature, is necessarily at enmity
with its parent. Unless the arguments set forth
in the Prolegomena, in the simplest language at
my command, have some flaw which I am unable
to discern, this seeming paradox is a truth, as
great as it is plain, the recognition of which
is fundamental for the ethical philosopher.

We cannot do without our inheritance from the
forefathers who were the puppets of the cosmic
process; the society which renounces it must
be destroyed from without. Still less can we do
with too much of it; the society in which it
dominates must be destroyed from within.

The motive of the drama of human life is the necessity, laid upon every man who comes into the world, of discovering the mean between self-assertion and self-restraint suited to his character and his circumstances. And the eternally tragic aspect of the drama lies in this: that the problem set before us is one the elements of which can be but imperfectly known, and of which even an approximately right solution rarely presents itself, until that stern critic, aged experience, has been furnished with ample justification for venting his sarcastic humour upon the irreparable blunders we have already made.

CONTENTS

I

EVOLUTION AND ETHICS

PROLEGOMENA

[1894]

I

IT may be safely assumed that, two thousand
years ago, before Cæsar set foot in southern
Britain, the whole country-side visible from the
windows of the room in which I write, was in
what is called " the state of nature." Except, it may
be, by raising a few sepulchral mounds, such as
those which still, here and there, break the flowing
contours of the downs, man's hands had made no
mark upon it; and the thin veil of vegetation
which overspread the broad-backed heights and
the shelving sides of the coombs was unaffected
by his industry. The native grasses and weeds,
the scattered patches of gorse, contended with one
another for the possession of the scanty surface
soil; they fought against the droughts of summer,

the frosts of winter, and the furious gales which swept, with unbroken force, now from the Atlantic, and now from the North Sea, at all times of the year; they filled up, as they best might, the gaps made in their ranks by all sorts of underground and overground animal ravagers. One year with another, an average population, the floating balance of the unceasing struggle for existence among the indigenous plants, maintained itself. It is as little to be doubted, that an essentially similar state of nature prevailed, in this region, for many thousand years before the coming of Cæsar; and there is no assignable reason for denying that it might continue to exist through an equally prolonged futurity, except for the intervention of man.

Reckoned by our customary standards of duration, the native vegetation, like the " everlasting hills " which it clothes, seems a type of permanence. The little Amarella Gentians, which abound in some places to-day, are the descendants of those that were trodden underfoot by the prehistoric savages who have left their flint tools about, here and there; and they followed ancestors which, in the climate of the glacial epoch, probably flourished better than they do now. Compared with the long past of this humble plant, all the history of civilized men is but an episode.

Yet nothing is more certain than that, measured by the liberal scale of time-keeping of the universe, this present state of nature, however it may seem

to have gone and to go on for ever, is but a
fleeting phase of her infinite variety; merely the
last of the series of changes which the earth's surface has undergone in the course of the millions of
years of its existence. Turn back a square foot of
the thin turf, and the solid foundation of the land,
exposed in cliffs of chalk five hundred feet high on
the adjacent shore, yields full assurance of a time
when the sea covered the site of the " everlasting
hills "; and when the vegetation of what land lay
nearest, was as different from the present Flora of
the Sussex downs, as that of Central Africa now is.[1]
No less certain is it that, between the time during
which the chalk was formed and that at which the
original turf came into existence, thousands of
centuries elapsed, in the course of which, the state
of nature of the ages during which the chalk was
deposited, passed into that which now is, by
changes so slow that, in the coming and going of
the generations of men, had such witnessed them,
the contemporary conditions would have seemed
to be unchanging and unchangeable.

But it is also certain that, before the deposition
of the chalk, a vastly longer period had elapsed,
throughout which it is easy to follow the traces
of the same process of ceaseless modification and
of the internecine struggle for existence of living
things; and that even when we can get no further

---

[1] See "On a piece of Chalk" in the preceding volume of these
Essays (vol. viii. p. 1).

back, it is not because there is any reason to think we have reached the beginning, but because the trail of the most ancient life remains hidden, or has become obliterated.

Thus that state of nature of the world of plants, which we began by considering, is far from possessing the attribute of permanence. Rather its very essence is impermanence. It may have lasted twenty or thirty thousand years, it may last for twenty or thirty thousand years more, without obvious change; but, as surely as it has followed upon a very different state, so it will be followed by an equally different condition. That which endures is not one or another association of living forms, but the process of which the cosmos is the product, and of which these are among the transitory expressions. And in the living world, one of the most characteristic features of this cosmic process is the struggle for existence, the competition of each with all, the result of which is the selection, that is to say, the survival of those forms which, on the whole, are best adapted to the conditions which at any period obtain; and which are, therefore, in that respect, and only in that respect, the fittest.[1]   The acme reached by the cosmic process

---

[1] That every theory of evolution must be consistent not merely with progressive development, but with indefinite persistence in the same condition and with retrogressive modification, is a point which I have insisted upon repeatedly from the year 1862 till now. See *Collected Essays*, vol. ii. pp. 461-89 ; vol. iii. p. 33 ; vol. viii. p. 304.   In the address on "Geological

in the vegetation of the downs is seen in the turf, with its weeds and gorse. Under the conditions, they have come out of the struggle victorious; and, by surviving, have proved that they are the fittest to survive.

That the state of nature, at any time, is a temporary phase of a process of incessant change, which has been going on for innumerable ages, appears to me to be a proposition as well established as any in modern history. Paleontology assures us, in addition, that the ancient philosophers who, with less reason, held the same doctrine, erred in supposing that the phases formed a cycle, exactly repeating the past, exactly foreshadowing the future, in their rotations. On the contrary, it furnishes us with conclusive reasons for thinking that, if every link in the ancestry of these humble indigenous plants had been preserved and were accessible to us, the whole would present a converging series of forms of gradually diminishing complexity, until, at some period in the history of the earth, far more remote than any of which organic remains have yet been discovered, they would merge in those low groups among which the boundaries between animal and vegetable life become effaced.[1]

Contemporaneity and Persistent Types" (1862), the paleontological proofs of this proposition were, I believe, first set torth.

[1] "On the Border Territory between the Animal and the Vegetable Kingdoms," Essays, vol. viii. p. 162.

The word "evolution," now generally applied to the cosmic process, has had a singular history, and is used in various senses.[1] Taken in its popular signification it means progressive development, that is, gradual change from a condition of relative uniformity to one of relative complexity; but its connotation has been widened to include the phenomena of retrogressive metamorphosis, that is, of progress from a condition of relative complexity to one of relative uniformity.

As a natural process, of the same character as the development of a tree from its seed, or of a fowl from its egg, evolution excludes creation and all other kinds of supernatural intervention. As the expression of a fixed order, every stage of which is the effect of causes operating according to definite rules, the conception of evolution no less excludes that of chance. It is very desirable to remember that evolution is not an explanation of the cosmic process, but merely a generalized statement of the method and results of that process. And, further, that, if there is proof that the cosmic process was set going by any agent, then that agent will be the creator of it and of all its products, although supernatural intervention may remain strictly excluded from its further course.

So far as that limited revelation of the nature of things, which we call scientific knowledge, has

[1] See "Evolution in Biology," Essays, vol. ii. p. 187.

yet gone, it tends, with constantly increasing emphasis, to the belief that, not merely the world of plants, but that of animals; not merely living things, but the whole fabric of the earth; not merely our planet, but the whole solar system; not merely our star and its satellites, but the millions of similar bodies which bear witness to the order which pervades boundless space, and has endured through boundless time; are all working out their predestined courses of evolution.

With none of these have I anything to do, at present, except with that exhibited by the forms of life which tenant the earth. All plants and animals exhibit the tendency to vary, the causes of which have yet to be ascertained; it is the tendency of the conditions of life, at any given time, while favouring the existence of the variations best adapted to them, to oppose that of the rest and thus to exercise selection; and all living things tend to multiply without limit, while the means of support are limited; the obvious cause of which is the production of offspring more numerous than their progenitors, but with equal expectation of life in the actuarial sense. Without the first tendency there could be no evolution. Without the second, there would be no good reason why one variation should disappear and another take its place; that is to say there would be no selection. Without the third, the struggle for existence, the agent of the selective process in the state of nature, would vanish.[1]

Granting the existence of these tendencies, all the known facts of the history of plants and of animals may be brought into rational correlation. And this is more than can be said for any other hypothesis that I know of. Such hypotheses, for example, as that of the existence of a primitive, orderless chaos; of a passive and sluggish eternal matter moulded, with but partial success, by archetypal ideas; of a brand-new world-stuff suddenly created and swiftly shaped by a supernatural power; receive no encouragement, but the contrary, from our present knowledge. That our earth may once have formed part of a nebulous cosmic magma is certainly possible, indeed seems highly probable; but there is no reason to doubt that order reigned there, as completely as amidst what we regard as the most finished works of nature or of man.[2] The faith which is born of knowledge, finds its object in an eternal order, bringing forth ceaseless change, through endless time, in endless space; the manifestations of the cosmic energy alternating between phases of potentiality and phases of explication. It may be that, as Kant suggests,[3] every cosmic

---

[1] *Collected Essays*, vol. ii. *passim.*
[2] *Ibid.*, vol. iv. p. 138 ; vol. v. pp. 71-73.
[3] *Ibid.*, vol. viii. p. 321.

magma predestined to evolve into a new world, has been the no less predestined end of a vanished predecessor.

II

Three or four years have elapsed since the state of nature, to which I have referred, was brought to an end, so far as a small patch of the soil is concerned, by the intervention of man. The patch was cut off from the rest by a wall; within the area thus protected, the native vegetation was, as far as possible, extirpated; while a colony of strange plants was imported and set down in its place. In short, it was made into a garden. At the present time, this artificially treated area presents an aspect extraordinarily different from that of so much of the land as remains in the state of nature, outside the wall. Trees, shrubs, and herbs, many of them appertaining to the state of nature of remote parts of the globe, abound and flourish. Moreover, considerable quantities of vegetables, fruits, and flowers are produced, of kinds which neither now exist, nor have ever existed, except under conditions such as obtain in the garden; and which, therefore, are as much works of the art of man as the frames and glass-houses in which some of them are raised. That the "state of Art," thus created in the state of nature by man, is sustained by and dependent on him, would at once become apparent, if the watchful supervision of the gardener were withdrawn, and the antagonistic influences of the general cosmic process were no longer sedulously warded off, or counteracted. The walls and gates would decay; quadrupedal and bipedal intruders would devour and tread down the useful and beautiful plants; birds, insects, blight, and mildew would work their will; the seeds of the native plants, carried by winds or other agencies, would immigrate, and in virtue of their long-earned special adaptation to the local conditions, these despised native weeds would soon choke their choice exotic rivals. A century or two hence, little beyond the foundations of the wall and of the houses and frames would be left, in evidence of the victory of the cosmic powers at work in the state of nature, over the temporary obstacles to their supremacy, set up by the art of the horticulturist.

It will be admitted that the garden is as much a work of art,[1] or artifice, as anything that can be mentioned. The energy localised in certain human bodies, directed by similarly localised intellects, has produced a collocation of other material bodies which could not be brought about in the state of nature. The same proposition is true of all the

---

[1] The sense of the term "Art" is becoming narrowed; "work of Art" to most people means a picture, a statue, or a piece of *bijouterie*; by way of compensation "artist" has included in its wide embrace cooks and ballet girls, no less than painters and sculptors.

works of man's hands, from a flint implement to a cathedral or a chronometer; and it is because it is true, that we call these things artificial, term them works of art, or artifice, by way of distinguishing them from the products of the cosmic process, working outside man, which we call natural, or works of nature. The distinction thus drawn between the works of nature and those of man, is universally recognised; and it is, as I conceive, both useful and justifiable.

III

No doubt, it may be properly urged that the operation of human energy and intelligence, which has brought into existence and maintains the garden, by what I have called "the horticultural process," is, strictly speaking, part and parcel of the cosmic process. And no one could more readily agree to that proposition than I. In fact, I do not know that any one has taken more pains than I have, during the last thirty years, to insist upon the doctrine, so much reviled in the early part of that period, that man, physical, intellectual, and moral, is as much a part of nature, as purely a product of the cosmic process, as the humblest weed.[1]

But if, following up this admission, it is urged

[1] See "Man's Place in Nature," *Collected Essays*, vol. vii., and "On the Struggle for Existence in Human Society"(1888), below.

that, such being the case, the cosmic process cannot be in antagonism with that horticultural process which is part of itself—I can only reply, that if the conclusion that the two are antagonistic is logically absurd, I am sorry for logic, because, as we have seen, the fact is so. The garden is in the same position as every other work of man's art; it is a result of the cosmic process working through and by human energy and intelligence ; and, as is the case with every other artificial thing set up in the state of nature, the influences of the latter are constantly tending to break it down and destroy it. No doubt, the Forth bridge and an ironclad in the offing, are, in ultimate resort, products of the cosmic process ; as much so as the river which flows under the one, or the seawater on which the other floats. Nevertheless, every breeze strains the bridge a little, every tide does something to weaken its foundations ; every change of temperature alters the adjustment of its parts, produces friction and consequent wear and tear. From time to time, the bridge must be repaired, just as the ironclad must go into dock ; simply because nature is always tending to reclaim that which her child, man, has borrowed from her and has arranged in combinations which are not those favoured by the general cosmic process.

Thus, it is not only true that the cosmic energy, working through man upon a portion of the plant world, opposes the same energy as it works through the state of nature, but a similar antagonism is everywhere manifest between the artificial and the natural. Even in the state of nature itself, what is the struggle for existence but the antagonism of the results of the cosmic process in the region of life, one to another? [1]

IV

Not only is the state of nature hostile to the state of art of the garden; but the principle of the horticultural process, by which the latter is created and maintained, is antithetic to that of the cosmic process. The characteristic feature of the latter is the intense and unceasing competition of the struggle for existence. The characteristic of the former is the elimination of that struggle, by the removal of the conditions which give rise to it. The tendency of the cosmic process is to bring about the adjustment of the forms of plant life to the current conditions; the tendency of the horticultural process is the adjustment of the conditions to the needs of the forms of plant life which the gardener desires to raise.

The cosmic process uses unrestricted multiplica-

[1] Or to put the case still more simply. When a man lays hold of the two ends of a piece of string and pulls them, with intent to break it, the right arm is certainly exerted in antagonism to the left arm ; yet both arms derive their energy from the same original source.

tion as the means whereby hundreds compete for the place and nourishment adequate for one; it employs frost and drought to cut off the weak and unfortunate; to survive, there is need not only of strength, but of flexibility and of good fortune.

The gardener, on the other hand, restricts multiplication; provides that each plant shall have sufficient space and nourishment; protects from frost and drought; and, in every other way attempts to modify the conditions, in such a manner as to bring about the survival of those forms which most nearly approach the standard of the useful, or the beautiful, which he has in his mind.

If the fruits and the tubers, the foliage and the flowers thus obtained, reach, or sufficiently approach, that ideal, there is no reason why the *status quo* attained should not be indefinitely prolonged. So long as the state of nature remains approximately the same, so long will the energy and intelligence which created the garden suffice to maintain it. However, the limits within which this mastery of man over nature can be maintained are narrow. If the conditions of the cretaceous epoch returned, I fear the most skilful of gardeners would have to give up the cultivation of apples and gooseberries; while, if those of the glacial period once again obtained, open asparagus beds would be superfluous, and the training of fruit trees against the most favourable of south walls, a waste of time and trouble.

But it is extremely important to note that, the state of nature remaining the same, if the produce does not satisfy the gardener, it may be made to approach his ideal more closely. Although the struggle for existence may be at end, the possibility of progress remains. In discussions on these topics, it is often strangely forgotten that the essential conditions of the modification, or evolution, of living things are variation and hereditary transmission. Selection is the means by which certain variations are favoured and their progeny preserved. But the struggle for existence is only one of the means by which selection may be effected. The endless varieties of cultivated flowers, fruits, roots, tubers, and bulbs are not products of selection by means of the struggle for existence, but of direct selection, in view of an ideal of utility or beauty. Amidst a multitude of plants, occupying the same station and subjected to the same conditions, in the garden, varieties arise. The varieties tending in a given direction are preserved, and the rest are destroyed. And the same process takes place among the varieties until, for example, the wild kale becomes a cabbage, or the wild *Viola tricolor* a prize pansy.

V

The process of colonization presents analogies
to the formation of a garden which are highly
instructive. Suppose a shipload of English
colonists sent to form a settlement, in such a
country as Tasmania was in the middle of the last
century. On landing, they find themselves in the
midst of a state of nature, widely different from
that left behind them in everything but the most
general physical conditions. The common plants,
the common birds and quadrupeds, are as totally
distinct as the men from anything to be seen on
the side of the globe from which they come.
The colonists proceed to put an end to this state
of things over as large an area as they desire to
occupy. They clear away the native vegetation,
extirpate or drive out the animal population, so
far as may be necessary, and take measures to
defend themselves from the re-immigration of
either. In their place, they introduce English
grain and fruit trees; English dogs, sheep, cattle,
horses; and English men; in fact, they set up a
new Flora and Fauna and a new variety of mankind,
within the old state of nature. Their farms and
pastures represent a garden on a great scale, and
themselves the gardeners who have to keep it up,
in watchful antagonism to the old *régime*. Considered as a whole, the colony is a composite unit
introduced into the old state of nature; and,

thenceforward, a competitor in the struggle for existence, to conquer or be vanquished.

Under the conditions supposed, there is no doubt of the result, if the work of the colonists be carried out energetically and with intelligent combination of all their forces. On the other hand, if they are slothful, stupid, and careless; or if they waste their energies in contests with one another, the chances are that the old state of nature will have the best of it. The native savage will destroy the immigrant civilized man; of the English animals and plants some will be extirpated by their indigenous rivals, others will pass into the feral state and themselves become components of the state of nature. In a few decades, all other traces of the settlement will have vanished.

VI

Let us now imagine that some administrative authority, as far superior in power and intelligence to men, as men are to their cattle, is set over the colony, charged to deal with its human elements in such a manner as to assure the victory of the settlement over the antagonistic influences of the state of nature in which it is set down. He would proceed in the same fashion as that in which the gardener dealt with his garden. In the first place, he would, as far as possible, put a

stop to the influence of external competition by
thoroughly extirpating and excluding the native
rivals, whether men, beasts, or plants. And
our administrator would select his human agents,
with a view to his ideal of a successful colony,
just as the gardener selects his plants with a view
to his ideal of useful or beautiful products.

In the second place, in order that no struggle
for the means of existence between these human
agents should weaken the efficiency of the corporate whole in the battle with the state of
nature, he would make arrangements by which
each would be provided with those means; and
would be relieved from the fear of being deprived
of them by his stronger or more cunning fellows.
Laws, sanctioned by the combined force of the
colony, would restrain the self-assertion of each
man within the limits required for the maintenance of peace. In other words, the cosmic struggle
for existence, as between man and man, would be
rigorously suppressed; and selection, by its means,
would be as completely excluded as it is from
the garden.

At the same time, the obstacles to the full
development of the capacities of the colonists
by other conditions of the state of nature
than those already mentioned, would be re-
moved by the creation of artificial conditions of
existence of a more favourable character. Protection against extremes of heat and cold would

be afforded by houses and clothing; drainage
and irrigation works would antagonise the effects
of excessive rain and excessive drought; roads,
bridges, canals, carriages, and ships would over-
come the natural obstacles to locomotion and
transport; mechanical engines would supple-
ment the natural strength of men and of
their draught animals; hygienic precautions
would check, or remove, the natural causes of
disease. With every step of this progress in
civilization, the colonists would become more and
more independent of the state of nature; more
and more, their lives would be conditioned by a
state of art. In order to attain his ends, the administrator would have to avail himself of the
courage, industry, and co-operative intelligence of
the settlers; and it is plain that the interest of
the community would be best served by increasing the proportion of persons who possess such
qualities, and diminishing that of persons devoid
of them. In other words, by selection directed
towards an ideal.

Thus the administrator might look to the
establishment of an earthly paradise, a true
garden of Eden, in which all things should
work together towards the well-being of the
gardeners: within which the cosmic process,
the coarse struggle for existence of the state
of nature, should be abolished; in which that
state should be replaced by a state of art;

where every plant and every lower animal should be adapted to human wants, and would perish if human supervision and protection were withdrawn; where men themselves should have been selected, with a view to their efficiency as organs for the performance of the functions of a perfected society. And this ideal polity would have been brought about, not by gradually adjusting the men to the conditions around them, but by creating artificial conditions for them; not by allowing the free play of the struggle for existence, but by excluding that struggle; and by substituting selection directed towards the administrator's ideal for the selection it exercises.

VII

But the Eden would have its serpent, and a very subtle beast too. Man shares with the rest of the living world the mighty instinct of reproduction and its consequence, the tendency to multiply with great rapidity. The better the measures of the administrator achieved their object, the more completely the destructive agencies of the state of nature were defeated, the less would that multiplication be checked.

On the other hand, within the colony, the enforcement of peace, which deprives every man of the power to take away the means of existence from another, simply because he is the stronger

would have put an end to the struggle for existence between the colonists, and the competition for
the commodities of existence, which would alone
remain, is no check upon population.

Thus, as soon as the colonists began to multiply,
the administrator would have to face the tendency
to the reintroduction of the cosmic struggle into
his artificial fabric, in consequence of the competition, not merely for the commodities, but for the
means of existence. When the colony reached
the limit of possible expansion, the surplus population must be disposed of somehow; or the fierce
struggle for existence must recommence and
destroy that peace, which is the fundamental condition of the maintenance of the state of art
against the state of nature.

Supposing the administrator to be guided by
purely scientific considerations, he would, like the
gardener, meet this most serious difficulty by
systematic extirpation, or exclusion, of the superfluous. The hopelessly diseased, the infirm aged,
the weak or deformed in body or in mind, the
excess of infants born, would be put away, as the
gardener pulls up defective and superfluous plants,
or the breeder destroys undesirable cattle. Only
the strong and the healthy, carefully matched, with
a view to the progeny best adapted to the purposes of the administrator, would be permitted to
perpetuate their kind.

VIII

Of the more thoroughgoing of the multitudinous attempts to apply the principles of cosmic evolution, or what are supposed to be such, to social and political problems, which have appeared of late years, a considerable proportion appear to me to be based upon the notion that human society is competent to furnish, from its own resources, an administrator of the kind I have imagined. The pigeons, in short, are to be their own Sir John Sebright.[1] A despotic government, whether individual or collective, is to be endowed with the preternatural intelligence, and with what, I am afraid, many will consider the preternatural ruthlessness, required for the purpose of carrying out the principle of improvement by selection, with the somewhat drastic thoroughness upon which the success of the method depends. Experience certainly does not justify us in limiting the ruthlessness of individual "saviours of society"; and, on the well-known grounds of the aphorism which denies both body and soul to corporations, it seems probable (indeed the belief is not without support in history) that a collective despotism, a mob got to believe in its own divine right by demagogic missionaries, would be capable of more thorough

---

[1] Not that the conception of such a society is necessarily based upon the idea of evolution. The Platonic state testifies to the contrary.

work in this direction than any single tyrant, puffed up with the same illusion, has ever achieved. But intelligence is another affair. The fact that "saviours of society" take to that trade is evidence enough that they have none to spare. And such as they possess is generally sold to the capitalists of physical force on whose resources they depend. However, I doubt whether even the keenest judge of character, if he had before him a hundred boys and girls under fourteen, could pick out, with the least chance of success, those who should be kept, as certain to be serviceable members of the polity, and those who should be chloroformed, as equally sure to be stupid, idle, or vicious. The "points" of a good or of a bad citizen are really far harder to discern than those of a puppy or a short-horn calf; many do not show themselves before the practical difficulties of life stimulate manhood to full exertion. And by that time the mischief is done. The evil stock, if it be one, has had time to multiply, and selection is nullified.

IX

I have other reasons for fearing that this logical ideal of evolutionary regimentation—this pigeon-fanciers' polity—is unattainable. In the absence of any such a severely scientific adminis-trator as we have been dreaming of, human society is kept together by bonds of such a singular character, that the attempt to perfect society after his fashion would run serious risk of loosening them.

Social organization is not peculiar to men. Other societies, such as those constituted by bees and ants, have also arisen out of the advantage of co-operation in the struggle for existence; and their resemblances to, and their differences from, human society are alike instructive. The society formed by the hive bee fulfils the ideal of the communistic aphorism "to each according to his needs, from each according to his capacity." Within it, the struggle for existence is strictly limited. Queen, drones, and workers have each their allotted sufficiency of food; each performs the function assigned to it in the economy of the hive, and all contribute to the success of the whole co-operative society in its competition with rival collectors of nectar and pollen and with other enemies, in the state of nature without. In the same sense as the garden, or the colony, is a work of human art, the bee polity is a work of apiarian art, brought about by the cosmic process, working through the organization of the hymenopterous type.

Now this society is the direct product of an organic necessity, impelling every member of it to a course of action which tends to the good of the whole. Each bee has its duty and none has any rights. Whether bees are susceptible of feeling and capable of thought is a question which cannot be dogmatically answered. As a pious opinion, I am disposed to deny them more than the merest rudiments of consciousness.[1] But it is curious to reflect that a thoughtful drone (workers and queens would have no leisure for speculation) with a turn for ethical philosophy, must needs profess himself an intuitive moralist of the purest water. He would point out, with perfect justice, that the devotion of the workers to a life of ceaseless toil for a mere subsistence wage, cannot be accounted for either by enlightened selfishness, or by any other sort of utilitarian motives; since these bees begin to work, without experience or reflection, as they emerge from the cell in which they are hatched. Plainly, an eternal and immutable principle, innate in each bee, can alone account for the phenomena. On the other hand, the biologist, who traces out all the extant stages of gradation between solitary and hive bees, as clearly sees in the latter, simply the perfection of an automatic mechanism, hammered out by the blows of the struggle for existence upon the progeny of the former, during long ages of constant variation.

[1] *Collected Essays*, vol. i., "Animal Automatism"; vol. v., "Prologue," pp. 45 *et seq.*

X

I see no reason to doubt that, at its origin, human society was as much a product of organic necessity as that of the bees.[1] The human family, to begin with, rested upon exactly the same conditions as those which gave rise to similar associations among animals lower in the scale. Further, it is easy to see that every increase in the duration of the family ties, with the resulting co-operation of a larger and larger number of descendants for protection and defence, would give the families in which such modification took place a distinct advantage over the others. And, as in the hive, the progressive limitation of the struggle for existence between the members of the family would involve increasing efficiency as regards outside competition.

But there is this vast and fundamental difference between bee society and human society. In the former, the members of the society are each organically predestined to the performance of one particular class of functions only. If they were endowed with desires, each could desire to perform none but those offices for which its organization specially fits it; and which, in view of the good of the whole, it is proper it should do. So long as a new queen does not make her appearance, rivalries and competition are absent from the bee polity.

[1] *Collected Essays*, vol. v., Prologue, pp. 50-54.

Among mankind, on the contrary, there is no such predestination to a sharply defined place in the social organism. However much men may differ in the quality of their intellects, the intensity of their passions, and the delicacy of their sensations, it cannot be said that one is fitted by his organization to be an agricultural labourer and nothing else, and another to be a landowner and nothing else. Moreover, with all their enormous differences in natural endowment, men agree in one thing, and that is their innate desire to enjoy the pleasures and to escape the pains of life; and, in short, to do nothing but that which it pleases them to do, without the least reference to the welfare of the society into which they are born. That is their inheritance (the reality at the bottom of the doctrine of original sin) from the long series of ancestors, human and semi-human and brutal, in whom the strength of this innate tendency to self-assertion was the condition of victory in the struggle for existence. That is the reason of the *aviditas vitæ* [1]—the insatiable hunger for enjoyment—of all mankind, which is one of the essential conditions of success in the war with the state of nature outside; and yet the sure agent of the destruction of society if allowed free play within.

The check upon this free play of self-assertion, or natural liberty, which is the necessary condition for the origin of human society, is the product

[1] See below. Romanes' Lecture, note 7.

of organic necessities of a different kind from
those upon which the constitution of the hive
depends. One of these is the mutual affection
of parent and offspring, intensified by the long
infancy of the human species. But the most
important is the tendency, so strongly
developed in man, to reproduce in himself actions and feelings similar to, or correlated with,
those of other men. Man is the most con-
summate of all mimics in the animal world;
none but himself can draw or model; none comes
near him in the scope, variety, and exactness of
vocal imitation; none is such a master of gesture;
while he seems to be impelled thus to imitate
for the pure pleasure of it. And there is
no such another emotional chameleon. By a
purely reflex operation of the mind, we take
the hue of passion of those who are about us,
or, it may be, the complementary colour. It is
not by any conscious "putting one's self in the
place" of a joyful or a suffering person that the
state of mind we call sympathy usually arises;[1]
indeed, it is often contrary to one's sense of

[1] Adam Smith makes the pithy observation that the man
who sympathises with a woman in childbed, cannot be said
to put himself in her place. ("The Theory of the Moral Sentiments," Part vii. sec. iii. chap. i.) Perhaps there is more
humour than force in the example; and, in spite of this
and other observations of the same tenor, I think that the
one defect of the remarkable work in which it occurs is that
it lays too much stress on conscious substitution, too little
on purely reflex sympathy.

right, and in spite of one's will, that "fellow-feeling makes us wondrous kind," or the reverse. However complete may be the indifference to public opinion, in a cool, intellectual view, of the traditional sage, it has not yet been my fortune to meet with any actual sage who took its hostile manifestations with entire equanimity. Indeed, I doubt if the philosopher lives, or ever has lived, who could know himself to be heartily despised by a street boy without some irritation. And, though one cannot justify Haman for wishing to hang Mordecai on such a very high gibbet, yet, really, the consciousness of the Vizier of Ahasuerus, as he went in and out of the gate, that this obscure Jew had no respect for him, must have been very annoying.[1]

It is needful only to look around us, to see that the greatest restrainer of the anti-social tendencies of men is fear, not of the law, but of the opinion of their fellows. The conventions of honour bind men who break legal, moral, and religious bonds; and, while people endure the extremity of physical pain rather than part with life, shame drives the weakest to suicide.

Every forward step of social progress brings men

---

[1] Esther v. 9–13. " . . . but when Haman saw Mordecai in the king's gate, that he stood not up, nor moved for him, he was full of indignation against Mordecai. . . . And Haman told them of the glory of his riches. . . . and all the things wherein the king had promoted him. . . . Yet all this availeth me nothing, so long as I see Mordecai the Jew sitting at the king's gate." What a shrewd exposure of human weakness it is!

into closer relations with their fellows, and increases the importance of the pleasures and pains derived from sympathy. We judge the acts of others by our own sympathies, and we judge our own acts by the sympathies of others, every day and all day long, from childhood upwards, until associations, as indissoluble as those of language, are formed between certain acts and the feelings of approbation or disapprobation. It becomes impossible to imagine some acts without disapprobation, or others without approbation of the actor, whether he be one's self, or any one else. We come to think in the acquired dialect of morals. An artificial personality, the "man within," as Adam Smith [1] calls conscience, is built up beside the natural personality. He is the watchman of society, charged to restrain the anti-social tendencies of the natural man within the limits required by social welfare.

XI

I have termed this evolution of the feelings out of which the primitive bonds of human society are so largely forged, into the organized and personified sympathy we call conscience, the ethical process.[2] So far as it tends to

[1] "Theory of the Moral Sentiments," Part iii. chap. 3. *On the influence and authority of conscience.*

[2] Worked out, in its essential features, chiefly by Hartley and Adam Smith, long before the modern doctrine of evolution was thought of. See *Note* below, p. 45.

make any human society more efficient in the struggle for existence with the state of nature, or with other societies, it works in harmonious contrast with the cosmic process. But it is none the less true that, since law and morals are restraints upon the struggle for existence between men in society, the ethical process is in opposition to the principle of the cosmic process, and tends to the suppression of the qualities best fitted for success in that struggle.[1]

It is further to be observed that, just as the self-assertion, necessary to the maintenance of society against the state of nature, will destroy that society if it is allowed free operation within; so the self-restraint, the essence of the ethical process, which is no less an essential condition of the existence of every polity, may, by excess, become ruinous to it.

Moralists of all ages and of all faiths, attending only to the relations of men towards one another in an ideal society, have agreed upon the "golden rule," "Do as you would be done by." In other words, let sympathy be your guide; put yourself in the place of the man towards whom your action is directed; and do to him what you would like to have done to yourself under the circumstances. However much one may admire the generosity of such a rule of con-

---

[1] See the essay "On the Struggle for Existence in Human Society" below; and *Collected Essays*, vol. i. p. 276, for Kant's recognition of these facts.

duct; however confident one may be that average men may be thoroughly depended upon not to carry it out to its full logical consequences; it is nevertheless desirable to recognise the fact that these consequences are incompatible with the existence of a civil state, under any circumstances of this world which have obtained, or, so far as one can see, are, likely to come to pass.

For I imagine there can be no doubt that the great desire of every wrongdoer is to escape from the painful consequences of his actions. If I put myself in the place of the man who has robbed me, I find that I am possessed by an exceeding desire not to be fined or imprisoned; if in that of the man who has smitten me on one cheek, I contemplate with satisfaction the absence of any worse result than the turning of the other cheek for like treatment. Strictly observed, the " golden rule " involves the negation of law by the refusal to put it in motion against law-breakers; and, as regards the external relations of a polity, it is the refusal to continue the struggle for existence. It can be obeyed, even partially, only under the protection of a society which repudiates it. Without such shelter, the followers of the " golden rule " may indulge in hopes of heaven, but they must reckon with the certainty that other people will be masters of the earth.

What would become of the garden if the gardener treated all the weeds and slugs and birds and trespassers as he would like to be treated, if he were in their place ?

XII

Under the preceding heads, I have endeavoured to represent in broad, but I hope faithful, outlines the essential features of the state of nature and of that cosmic process of which it is the outcome, so far as was needful for my argument ; I have contrasted with the state of nature the state of art, produced by human intelligence and energy, as it is exemplified by a garden ; and I have shown that the state of art, here and elsewhere, can be maintained only by the constant counteraction of the hostile influences of the state of nature. Further, I have pointed out that the " horticultural process " which thus sets itself against the " cosmic process " is opposed to the latter in principle, in so far as it tends to arrest the struggle for existence, by restraining the multiplication which is one of the chief causes of that struggle, and by creating artificial conditions of life, better adapted to the cultivated plants than are the conditions of the state of nature. And I have dwelt upon the fact that, though the progressive modification, which is the consequence of the struggle for existence in the state of nature, is at an end, such modification may still be effected by that selection, in view of an ideal of usefulness, or of pleasantness, to man, of which the state of nature knows nothing.

I have proceeded to show that a colony, set down in a country in the state of nature, presents close analogies with a garden; and I have indicated the course of action which an administrator, able and willing to carry out horticultural principles, would adopt, in order to secure the success of such a newly formed polity, supposing it to be capable of indefinite expansion. In the contrary case, I have shown that difficulties must arise; that the unlimited increase of the population over a limited area must, sooner or later, reintroduce into the colony that struggle for the means of existence between the colonists, which it was the primary object of the administrator to exclude, insomuch as it is fatal to the mutual peace which is the prime condition of the union of men in society.

I have briefly described the nature of the only radical cure, known to me, for the disease which would thus threaten the existence of the colony; and, however regretfully, I have been obliged to admit that this rigorously scientific method of applying the principles of evolution to human society hardly comes within the region of practical politics; not for want of will on the part of a great many people; but because, for one reason, there is no hope that mere human beings will ever possess enough intelligence to select the fittest. And I have adduced other grounds for arriving at the same conclusion.

I have pointed out that human society took its rise in the organic necessities expressed by imitation and by the sympathetic emotions; and that, in the struggle for existence with the state of nature and with other societies, as part of it, those in which men were thus led to close co-operation had a great advantage.[1] But, since each man retained more or less of the faculties common to all the rest, and especially a full share of the desire for unlimited self-gratification, the struggle for existence within society could only be gradually eliminated. So long as any of it remained, society continued to be an imperfect instrument of the struggle for existence and, consequently, was improvable by the selective influence of that struggle. Other things being alike, the tribe of savages in which order was best maintained; in which there was most security within the tribe and the most loyal mutual support outside it, would be the survivors.

I have termed this gradual strengthening of the social bond, which, though it arrests the struggle for existence inside society, up to a certain point improves the chances of society, as a corporate whole, in the cosmic struggle—the ethical process. I have endeavoured to show that, when the ethical process has advanced so

---

[1] *Collected Essays*, vol. v., Prologue, p. 52.

far as to secure every member of the society in the possession of the means of existence, the struggle for existence, as between man and man, within that society is, *ipso facto*, at an end. And, as it is undeniable that the most highly civilized societies have substantially reached this position, it follows that, so far as they are concerned, the struggle for existence can play no important part within them.[1] In other words, the kind of evolution which is brought about in the state of nature cannot take place.

I have further shown cause for the belief that direct selection, after the fashion of the horticulturist and the breeder, neither has played, nor can play, any important part in the evolution of society; apart from other reasons, because I do not see how such selection could be practised without a serious weakening, it may be the destruction, of the bonds which hold society together. It strikes me that men who are accustomed to contemplate the active or passive extirpation of the weak, the unfortunate, and the superfluous; who justify that conduct on the ground that it has the sanction of the cosmic process, and is the only way of ensuring the progress of the race; who, if

---

[1] Whether the struggle for existence with the state of nature and with other societies, so far as they stand in the relation of the state of nature with it, exerts a selective influence upon modern society, and in what direction, are questions not easy to answer. The problem of the effect of military and industrial warfare upon those who wage it is very complicated.

they are consistent, must rank medicine among the black arts and count the physician a mischievous preserver of the unfit; on whose matrimonial undertakings the principles of the stud have the chief influence; whose whole lives, therefore, are an education in the noble art of suppressing natural affection and sympathy, are not likely to have any large stock of these commodities left. But, without them, there is no conscience, nor any restraint on the conduct of men, except the calculation of self-interest, the balancing of certain present gratifications against doubtful future pains; and experience tells us how much that is worth. Every day, we see firm believers in the hell of the theologians commit acts by which, as they believe when cool, they risk eternal punishment; while they hold back from those which are opposed to the sympathies of their associates.

XIII

That progressive modification of civilization which passes by the name of the "evolution of society," is, in fact, a process of an essentially different character, both from that which brings about the evolution of species, in the state of nature, and from that which gives rise to the evolution of varieties, in the state of art.

There can be no doubt that vast changes have taken place in English civilization since the reign of the Tudors. But I am not aware of a particle of evidence in favour of the conclusion that this evolutionary process has been accompanied by any modification of the physical, or the mental, characters of the men who have been the subjects of it. I have not met with any grounds for suspecting that the average Englishmen of to-day are sensibly different from those that Shakspere knew and drew. We look into his magic mirror of the Elizabethan age, and behold, nowise darkly, the presentment of ourselves.

During these three centuries, from the reign of Elizabeth to that of Victoria, the struggle for existence between man and man has been so largely restrained among the great mass of the population (except for one or two short intervals of civil war), that it can have had little, or no, selective operation. As to anything comparable to direct selection, it has been practised on so small a scale that it may also be neglected. The criminal law, in so far as by putting to death, or by subjecting to long periods of imprisonment, those who infringe its provisions, it prevents the propagation of hereditary criminal tendencies; and the poor-law, in so far as it separates married couples, whose destitution arises from hereditary defects of character, are doubtless selective agents operating in favour of the non-criminal and the more effective members of society. But the proportion of the population which they influence is very small; and, generally, the hereditary criminal and the hereditary pauper have propagated their kind before the law affects them. In a large proportion of cases, crime and pauperism have nothing to do with heredity; but are the consequence, partly, of circumstances and, partly, of the possession of qualities, which, under different conditions of life, might have excited esteem and even admiration. It was a shrewd man of the world who, in discussing sewage problems, remarked that dirt is riches in the wrong place; and that sound aphorism has moral applications. The benevolence and open-handed generosity which adorn a rich man, may make a pauper of a poor one; the energy and courage to which the successful soldier owes his rise, the cool and daring subtlety to which the great financier owes his fortune, may very easily, under unfavourable conditions, lead their possessors to the gallows, or to the hulks. Moreover, it is fairly probable that the children of a 'failure' will receive from their other parent just that little modification of character which makes all the difference. I sometimes wonder whether people, who talk so freely about extirpating the unfit, ever dispassionately consider their own history. Surely, one must be very 'fit,' indeed, not to know of an occasion, or perhaps two, in one's life, when it would have been only too easy to qualify for a place among the 'unfit.'

In my belief the innate qualities, physical, intellectual, and moral, of our nation have remained substantially the same for the last four or five centuries. If the struggle for existence has affected us to any serious extent (and I doubt it) it has been, indirectly, through our military and industrial wars with other nations.

XIV

What is often called the struggle for existence in society (I plead guilty to having used the term too loosely myself), is a contest, not for the means of existence, but for the means of enjoyment. Those who occupy the first places in this practical competitive examination are the rich and the influential; those who fail, more or less, occupy the lower places, down to the squalid obscurity of the pauper and the criminal. Upon the most liberal estimate, I suppose the former group will not amount to two per cent. of the population. I doubt if the latter exceeds another two per cent.; but let it be supposed, for the sake of argument, that it is as great as five per cent.[1]

As it is only in the latter group that anything comparable to the struggle for existence in the state of nature can take place; as it is only

---

[1] Those who read the last Essay in this volume will not accuse me of wishing to attenuate the evil of the existence of this group, whether great or small.

among this twentieth of the whole people that numerous men, women, and children die of rapid or slow starvation, or of the diseases incidental to permanently bad conditions of life ; and as there is nothing to prevent their multiplication before they are killed off, while, in spite of greater infant mortality, they increase faster than the rich ; it seems clear that the struggle for existence in this class can have no appreciable selective influence upon the other 95 per cent. of the population.

What sort of a sheep breeder would he be who should content himself with picking out the worst fifty out of a thousand, leaving them on a barren common till the weakest starved, and then letting the survivors go back to mix with the rest ? And the parallel is too favourable ; since in a large number of cases, the actual poor and the convicted criminals are neither the weakest nor the worst.

In the struggle for the means of enjoyment, the qualities which ensure success are energy, industry, intellectual capacity, tenacity of purpose, and, at least as much sympathy as is necessary to make a man understand the feelings of his fellows. Were there none of those artificial arrangements by which fools and knaves are kept at the top of society instead of sinking to their natural place at the bottom,[1] the struggle for the means of

1 I have elsewhere lamented the absence from society of

enjoyment would ensure a constant circulation of the human units of the social compound, from the bottom to the top and from the top to the bottom. The survivors of the contest, those who continued to form the great bulk of the polity, would not be those 'fittest' who got to the very top, but the great body of the moderately "fit," whose numbers and superior propagative power, enable them always to swamp the exceptionally endowed minority.

I think it must be obvious to every one, that, whether we consider the internal or the external interests of society, it is desirable they should be in the hands of those who are endowed with the largest share of energy, of industry, of intellectual capacity, of tenacity of purpose, while they are not devoid of sympathetic humanity; and, in so far as the struggle for the means of enjoyment tends to place such men in possession of wealth and influence, it is a process which tends to the good of society. But the process, as we have seen, has no real resemblance to that which adapts living beings to current conditions in the state of nature; nor any to the artificial selection of the horticulturist.

a machinery for facilitating the descent of incapacity. "Administrative Nihilism." *Collected Essays,* vol. i. p. 54.

XV

To return, once more, to the parallel of horticulture. In the modern world, the gardening of men by themselves is practically restricted to the performance, not of selection, but of that other function of the gardener, the creation of conditions more favourable than those of the state of nature ; to the end of facilitating the free expansion of the innate faculties of the citizen, so far as it is consistent with the general good. And the business of the moral and political philosopher appears to me to be the ascertainment, by the same method of observation, experiment, and ratiocination, as is practised in other kinds of scientific work, of the course of conduct which will best conduce to that end.

But, supposing this course of conduct to be scientifically determined and carefully followed out, it cannot put an end to the struggle for existence in the state of nature ; and it will not so much as tend, in any way, to the adaptation of man to that state. Even should the whole human race be absorbed in one vast polity, within which "absolute political justice" reigns, the struggle for existence with the state of nature outside it, and the tendency to the return of the struggle within, in consequence of over-multiplication, will remain ; and, unless men's inheritance from the ancestors who fought a good fight in the state of

nature, their dose of original sin, is rooted out by
some method at present unrevealed, at any rate
to disbelievers in supernaturalism, every child
born into the world will still bring with him the
instinct of unlimited self-assertion. He will have
to learn the lesson of self-restraint and renunciation. But the practice of self-restraint and renunciation is not happiness, though it may be
something much better.

That man, as a 'political animal,' is susceptible of a vast amount of improvement, by education, by instruction, and by the application of his
intelligence to the adaptation of the conditions
of life to his higher needs, I entertain not the
slightest doubt. But, so long as he remains liable
to error, intellectual or moral; so long as he is
compelled to be perpetually on guard against the
cosmic forces, whose ends are not his ends, without
and within himself; so long as he is haunted by
inexpugnable memories and hopeless aspirations;
so long as the recognition of his intellectual limita-
tions forces him to acknowledge his incapacity to
penetrate the mystery of existence; the prospect
of attaining untroubled happiness, or of a state
which can, even remotely, deserve the title of
perfection, appears to me to be as misleading an
illusion as ever was dangled before the eyes of poor
humanity. And there have been many of them.

That which lies before the human race is a
constant struggle to maintain and improve, in

opposition to the State of Nature, the State of Art of an organized polity; in which, and by which, man may develop a worthy civilization, capable of maintaining and constantly improving itself, until the evolution of our globe shall have entered so far upon its downward course that the cosmic process resumes its sway; and, once more, the State of Nature prevails over the surface of our planet.

Note (see p. 30).—It seems the fashion nowadays to ignore Hartley; though, a century and a half ago, he not only laid the foundations but built up much of the superstructure of a true theory of the Evolution of the intellectual and moral faculties. He speaks of what I have termed the ethical process as "our Progress from Self-interest to **Self-annihilation.**" *Observations on Man* (1749), vol. ii. p. 281.

II

EVOLUTION AND ETHICS

[*The Romanes Lecture*, 1893]

Soleo enim et in aliena castra transire, non tanquam transfuga
sed tanquam explorator. (L. Annæi Senecæ Epist. II. 4.)

THERE is a delightful child's story, known by
the title of " Jack and the Bean-stalk," with
which my contemporaries who are present will be
familiar. But so many of our grave and reverend
juniors have been brought up on severer intellectual diet, and, perhaps, have become acquainted
with fairyland only through primers of comparative
mythology, that it may be needful to give an outline of the tale. It is a legend of a bean-plant,
which grows and grows until it reaches the high
heavens and there spreads out into a vast canopy
of foliage. The hero, being moved to climb the
stalk, discovers that the leafy expanse supports a
world composed of the same elements as that
below, but yet strangely new; and his adventures
there, on which I may not dwell, must have com-

pletely changed his views of the nature of things; though the story, not having been composed by, or for, philosophers, has nothing to say about views.

My present enterprise has a certain analogy to that of the daring adventurer. I beg you to accompany me in an attempt to reach a world which, to many, is probably strange, by the help of a bean. It is, as you know, a simple, inert-looking thing. Yet, if planted under proper conditions, of which sufficient warmth is one of the most important, it manifests active powers of a very remarkable kind. A small green seedling emerges, rises to the surface of the soil, rapidly increases in size and, at the same time, undergoes a series of metamorphoses which do not excite our wonder as much as those which meet us in legendary history, merely because they are to be seen every day and all day long.

By insensible steps, the plant builds itself up into a large and various fabric of root, stem, leaves, flowers, and fruit, every one moulded within and without in accordance with an extremely complex but, at the same time, minutely defined pattern. In each of these complicated structures, as in their smallest constituents, there is an immanent energy which, in harmony with that resident in all the others, incessantly works towards the maintenance of the whole and the efficient performance of the part which it has to play in the economy of nature.

But no sooner has the edifice, reared with such exact elaboration, attained completeness, than it begins to crumble. By degrees, the plant withers and disappears from view, leaving behind more or fewer apparently inert and simple bodies, just like the bean from which it sprang; and, like it, endowed with the potentiality of giving rise to a similar cycle of manifestations.

Neither the poetic nor the scientific imagination is put to much strain in the search after analogies with this process of going forth and, as it were, returning to the starting-point. It may be likened to the ascent and descent of a slung stone, or the course of an arrow along its trajectory. Or we may say that the living energy takes first an upward and then a downward road. Or it may seem preferable to compare the expansion of the germ into the full-grown plant, to the unfolding of a fan, or to the rolling forth and widening of a stream; and thus to arrive at the conception of 'development,' or 'evolution.' Here as elsewhere, names are 'noise and smoke'; the important point is to have a clear and adequate conception of the fact signified by a name. And, in this case, the fact is the Sisyphæan process, in the course of which, the living and growing plant passes from the relative simplicity and latent potentiality of the seed to the full epiphany of a highly differentiated type, thence to fall back to simplicity and potentiality.

The value of a strong intellectual grasp of the
nature of this process lies in the circumstance that
what is true of the bean is true of living things in
general. From very low forms up to the highest
—in the animal no less than in the vegetable
kingdom—the process of life presents the same
appearance [1] of cyclical evolution. Nay, we have
but to cast our eyes over the rest of the world and
cyclical change presents itself on all sides. It
meets us in the water that flows to the sea and
returns to the springs; in the heavenly bodies
that wax and wane, go and return to their places;
in the inexorable sequence of the ages of man's
life; in that successive rise, apogee, and fall of
dynasties and of states which is the most pro-
minent topic of civil history.

As no man fording a swift stream can dip his
foot twice into the same water, so no man can,
with exactness, affirm of anything in the sensible
world that it is.[2] As he utters the words, nay,
as he thinks them, the predicate ceases to be
applicable; the present has become the past; the
'is' should be 'was.' And the more we learn of
the nature of things, the more evident is it that
what we call rest is only unperceived activity;
that seeming peace is silent but strenuous battle.
In every part, at every moment, the state of the
cosmos is the expression of a transitory adjustment of contending forces; a scene of strife, in
which all the combatants fall in turn. What is

true of each part, is true of the whole. Natural knowledge tends more and more to the conclusion that " all the choir of heaven and furniture of the earth " are the transitory forms of parcels of cosmic substance wending along the road of evolution, from nebulous potentiality, through endless growths of sun and planet and satellite; through all varieties of matter; through infinite diversities of life and thought; possibly, through modes of being of which we neither have a conception, nor are competent to form any, back to the indefinable latency from which they arose. Thus the most obvious attribute of the cosmos is its impermanence. It assumes the aspect not so much of a permanent entity as of a changeful process, in which naught endures save the flow of energy and the rational order which pervades it.

We have climbed our bean-stalk and have reached a wonderland in which the common and the familiar become things new and strange. In the exploration of the cosmic process thus typified, the highest intelligence of man finds inexhaustible employment; giants are subdued to our service; and the spiritual affections of the contemplative philosopher are engaged by beauties worthy of eternal constancy.

But there is another aspect of the cosmic process, so perfect as a mechanism, so beautiful as a work of art. Where the cosmopoietic energy works

through sentient beings, there arises, among its
other manifestations, that which we call pain or
suffering. This baleful product of evolution increases in quantity and in intensity, with advancing
grades of animal organization, until it attains its
highest level in man. Further, the consummation is not reached in man, the mere animal; nor
in man, the whole or half savage; but only in
man, the member of an organized polity. And
it is a necessary consequence of his attempt to live
in this way; that is, under those conditions which
are essential to the full development of his noblest
powers.

Man, the animal, in fact, has worked his way
to the headship of the sentient world, and has
become the superb animal which he is, in virtue
of his success in the struggle for existence. The
conditions having been of a certain order, man's
organization has adjusted itself to them better
than that of his competitors in the cosmic strife.
In the case of mankind, the self-assertion, the
unscrupulous seizing upon all that can be grasped,
the tenacious holding of all that can be kept,
which constitute the essence of the struggle for
existence, have answered. For his successful progress, throughout the savage state, man has been
largely indebted to those qualities which he shares
with the ape and the tiger; his exceptional
physical organization; his cunning, his sociability,
his curiosity, and his imitativeness; his ruthless

and ferocious destructiveness when his anger is roused by opposition.

But, in proportion as men have passed from anarchy to social organization, and in proportion as civilization has grown in worth, these deeply ingrained serviceable qualities have become defects. After the manner of successful persons, civilized man would gladly kick down the ladder by which he has climbed. He would be only too pleased to see 'the ape and tiger die.' But they decline to suit his convenience; and the unwelcome intrusion of these boon companions of his hot youth into the ranged existence of civil life adds pains and griefs, innumerable and immeasurably great, to those which the cosmic process necessarily brings on the mere animal. In fact, civilized man brands all these ape and tiger promptings with the name of sins; he punishes many of the acts which flow from them as crimes; and, in extreme cases, he does his best to put an end to the survival of the fittest of former days by axe and rope.

I have said that civilized man has reached this point; the assertion is perhaps too broad and general; I had better put it that ethical man has attained thereto. The science of ethics professes to furnish us with a reasoned rule of life; to tell us what is right action and why it is so. Whatever differences of opinion may exist among experts, there is a general consensus that the ape and tiger methods of the struggle for existence are **not** reconcilable with sound ethical principles.

The hero of our story descended the bean-stalk, and came back to the common world, where fare and work were alike hard ; where ugly competitors were much commoner than beautiful princesses ; and where the everlasting battle with self was much less sure to be crowned with victory than a turn-to with a giant. We have done the like. Thousands upon thousands of our fellows, thousands of years ago, have preceded us in finding themselves face to face with the same dread problem of evil. They also have seen that the cosmic process is evolution; that it is full of wonder, full of beauty, and, at the same time, full of pain. They have sought to discover the bearing of these great facts on ethics ; to find out whether there is, or is not, a sanction for morality in the ways of the cosmos.

Theories of the universe, in which the conception of evolution plays a leading part, were extant at least six centuries before our era. Certain knowledge of them, in the fifth century, reaches us from localities as distant as the valley of the Ganges and the Asiatic coasts of the Ægean. To the early philosophers of Hindostan, no less than to those of Ionia, the salient and characteristic feature of the phenomenal world was its changefulness; the unresting flow of all things, through birth to visible being and thence to not being, in which they could discern no sign of a beginning and for which they saw no prospect of an ending. It was no less plain to some of these antique fore-runners of modern philosophy that suffering is the badge of all the tribe of sentient things; that it is no accidental accompaniment, but an essential constituent of the cosmic process. The energetic Greek might find fierce joys in a world in which 'strife is father and king'; but the old Aryan spirit was subdued to quietism in the Indian sage; the mist of suffering which spread over humanity hid everything else from his view; to him life was one with suffering and suffering with life.

In Hindostan, as in Ionia, a period of relatively high and tolerably stable civilization had succeeded long ages of semi-barbarism and struggle. Out of wealth and security had come leisure and refine-ment, and, close at their heels, had followed the malady of thought. To the struggle for bare existence, which never ends, though it may be alleviated and partially disguised for a fortunate few, succeeded the struggle to make existence intelligible and to bring the order of things into harmony with the moral sense of man, which also never ends, but, for the thinking few, becomes keener with every increase of knowledge and with every step towards the realization of a worthy ideal of life.

Two thousand five hundred years ago, the value of civilization was as apparent as it is now; then, as now, it was obvious that only in the garden of an orderly polity can the finest fruits humanity is capable of bearing be produced. But it had also become evident that the blessings of culture were not unmixed. The garden was apt to turn into a hothouse. The stimulation of the senses, the pampering of the emotions, endlessly multiplied the sources of pleasure. The constant widening of the intellectual field indefinitely extended the range of that especially human faculty of looking before and after, which adds to the fleeting present those old and new worlds of the past and the future, wherein men dwell the more the higher their culture. But that very sharpening of the sense and that subtle refinement of emotion, which brought such a wealth of pleasures, were fatally attended by a proportional enlargement of the capacity for suffering; and the divine faculty of imagination, while it created new heavens and new earths, provided them with the corresponding hells of futile regret for the past and morbid anxiety for the future.[3] Finally, the inevitable penalty of over-stimulation, exhaustion, opened the gates of civilization to its great enemy, ennui; the stale and flat weariness when man delights not, nor woman neither; when all things are vanity and vexation; and life seems not worth living except to escape the bore of dying.

Even purely intellectual progress brings about its revenges. Problems settled in a rough and ready way by rude men, absorbed in action, demand renewed attention and show themselves to be still unread riddles when men have time to think. The beneficent demon, doubt, whose name is Legion and who dwells amongst the tombs of old faiths, enters into mankind and thenceforth refuses to be cast out. Sacred customs, venerable dooms of ancestral wisdom, hallowed by tradition and professing to hold good for all time, are put to the question. Cultured reflection asks for their credentials; judges them by its own standards; finally, gathers those of which it approves into ethical systems, in which the reasoning is rarely much more than a decent pretext for the adoption of foregone conclusions.

One of the oldest and most important elements in such systems is the conception of justice. Society is impossible unless those who are associated agree to observe certain rules of conduct towards one another; its stability depends on the steadiness with which they abide by that agreement; and, so far as they waver, that mutual trust which is the bond of society is weakened or destroyed. Wolves could not hunt in packs except for the real, though unexpressed, understanding that they should not attack one another during the chase. The most rudimentary polity is a pack of men living under the like tacit, or expressed, understanding; and having made the very important advance upon wolf society, that they agree to use the force of the whole body against individuals who violate it and in favour of those who observe it. This observance of a common understanding, with the consequent distribution of punishments and rewards according to accepted rules, received the name of justice, while the contrary was called injustice. Early ethics did not take much note of the animus of the violator of the rules. But civilization could not advance far, without the establishment of a capital distinction between the case of involuntary and that of wilful misdeed; between a merely wrong action and a guilty one. And, with increasing refinement of moral appreciation, the problem of desert, which arises out of this distinction, acquired more and more theoretical and practical importance. If life must be given for life, yet it was recognized that the unintentional slayer did not altogether deserve death; and, by a sort of compromise between the public and the private conception of justice, a sanctuary was provided in which he might take refuge from the avenger of blood.

The idea of justice thus underwent a gradual sublimation from punishment and reward according to acts, to punishment and reward according to desert; or, in other words, according to motive. Righteousness, that is, action from right motive, not only became synonymous with justice, but the positive constituent of innocence and the very heart of goodness.

Now when the ancient sage, whether Indian or Greek, who had attained to this conception of goodness, looked the world, and especially human life, in the face, he found it as hard as we do to bring the course of evolution into harmony with even the elementary requirements of the ethical ideal of the just and the good.

If there is one thing plainer than another, it is that neither the pleasures nor the pains of life, in the merely animal world, are distributed according to desert; for it is admittedly impossible for the lower orders of sentient beings to deserve either the one or the other. If there is a generalization from the facts of human life which has the assent of thoughtful men in every age and country, it is that the violator of ethical rules constantly escapes the punishment which he deserves; that the wicked flourishes like a green bay tree, while the righteous begs his bread; that the sins of the fathers are visited upon the children; that, in the realm of nature, ignorance is punished just as severely as wilful wrong; and that thousands upon thousands of innocent beings suffer for the crime, or the unintentional trespass of one.

Greek and Semite and Indian are agreed upon

this subject. The book of Job is at one with the
" Works and Days" and the Buddhist Sutras;
the Psalmist and the Preacher of Israel, with the
Tragic Poets of Greece. What is a more common
motive of the ancient tragedy in fact, than the
unfathomable injustice of the nature of things;
what is more deeply felt to be true than its pre-
sentation of the destruction of the blameless by
the work of his own hands, or by the fatal operation of the sins of others? Surely Œdipus was
pure of heart; it was the natural sequence of
events—the cosmic process—which drove him, in
all innocence, to slay his father and become the
husband of his mother, to the desolation of his
people and his own headlong ruin. Or to step, for
a moment, beyond the chronological limits I have
set myself, what constitutes the sempiternal at-
traction of Hamlet but the appeal to deepest
experience of that history of a no less blameless
dreamer, dragged, in spite of himself, into a world
out of joint; involved in a tangle of crime and
misery, created by one of the prime agents of the
cosmic process as it works in and through man?

Thus, brought before the tribunal of ethics, the
cosmos might well seem to stand condemned.
The conscience of man revolted against the moral
indifference of nature, and the microcosmic atom
should have found the illimitable macrocosm
guilty. But few, or none, ventured to record that
verdict.

In the great Semitic trial of this issue, Job takes refuge in silence and submission; the Indian and the Greek, less wise perhaps, attempt to reconcile the irreconcilable and plead for the defendant. To this end, the Greeks invented Theodicies; while the Indians devised what, in its ultimate form, must rather be termed a Cosmodicy. For, though Buddhism recognizes gods many and lords many, they are products of the cosmic process; and transitory, however long enduring, manifestations of its eternal activity. In the doctrine of transmigration, whatever its origin, Brahminical and Buddhist speculation found, ready to hand,[4] the means of constructing a plausible vindication of the ways of the cosmos to man. If this world is full of pain and sorrow; if grief and evil fall, like the rain, upon both the just and the unjust; it is because, like the rain, they are links in the endless chain of natural causation by which past, present, and future are indissolubly connected; and there is no more injustice in the one case than in the other. Every sentient being is reaping as it has sown; if not in this life, then in one or other of the infinite series of antecedent existences of which it is the latest term. The present distribution of good and evil is, therefore, the algebraical sum of accumulated positive and negative deserts; or, rather, it depends on the floating balance of the account. For it was not thought necessary that a complete settlement should ever take place. Arrears might stand over as a sort of 'hanging gale'; a period of celestial happiness just earned might be succeeded by ages of torment in a hideous nether world, the balance still overdue for some remote ancestral error.[5]

Whether the cosmic process looks any more moral than at first, after such a vindication, may perhaps be questioned. Yet this plea of justification is not less plausible than others; and none but very hasty thinkers will reject it on the ground of inherent absurdity. Like the doctrine of evolution itself, that of transmigration has its roots in the world of reality; and it may claim such support as the great argument from analogy is capable of supplying.

Everyday experience familiarizes us with the facts which are grouped under the name of heredity. Every one of us bears upon him obvious marks of his parentage, perhaps of remoter relationships. More particularly, the sum of tendencies to act in a certain way, which we call "character," is often to be traced through a long series of progenitors and collaterals. So we may justly say that this 'character'—this moral and intellectual essence of a man—does veritably pass over from one fleshly tabernacle to another, and does really transmigrate from generation to generation. In the new-born infant, the character of the stock lies latent, and the Ego is little more than a bundle of potentialities. But, very early, these become actualities; from childhood to age they manifest themselves in dulness or brightness, weakness or strength, viciousness or uprightness; and with each feature modified by confluence with another character, if by nothing else, the character passes on to its incarnation in new bodies.

The Indian philosophers called character, as thus defined, 'karma.'[6] It is this karma which passed from life to life and linked them in the chain of transmigrations; and they held that it is modified in each life, not merely by confluence of parentage, but by its own acts. They were, in fact, strong believers in the theory, so much disputed just at present, of the hereditary transmission of acquired characters. That the manifestation of the tendencies of a character may be greatly facilitated, or impeded, by conditions, of which self-discipline, or the absence of it, are among the most important, is indubitable; but that the character itself is modified in this way is by no means so certain; it is not so sure that the transmitted character of an evil liver is worse, or that of a righteous man better, than that which he received. Indian philosophy, however, did not admit of any doubt on this subject; the belief in the influence of conditions, notably of self-discipline, on the karma was not merely a necessary postulate of its theory of retribution, but it presented the only way of escape from the endless round of transmigrations.

The earlier forms of Indian philosophy agreed with those prevalent in our own times, in supposing the existence of a permanent reality, or ' substance,' beneath the shifting series of phenomena, whether of matter or of mind. The substance of the cosmos was ' Brahma,' that of the individual man ' Atman'; and the latter was separated from the former only, if I may so speak, by its phenomenal envelope, by the casing of sensations, thoughts and desires, pleasures and pains, which make up the illusive phantasmagoria of life. This the ignorant take for reality; their ' Atman' therefore remains eternally imprisoned in delusions, bound by the fetters of desire and scourged by the whip of misery. But the man who has attained enlightenment sees that the apparent reality is mere illusion, or, as was said a couple of thousand years later, that there is nothing good nor bad but thinking makes it so. If the cosmos " is just and of our pleasant vices makes instruments to scourge us," it would seem that the only way to escape from our heritage of evil is to destroy that fountain of desire whence our vices flow; to refuse any longer to be the instruments of the evolutionary process, and withdraw from the struggle for existence. If the karma is modifiable by self-discipline, if its coarser desires, one after another, can be extinguished, the ultimate fundamental desire of self-assertion, or the desire to be, may also be destroyed.[7]   Then the bubble of illusion will burst, and the freed individual 'Atman' will lose itself in the universal 'Brahma.'

Such seems to have been the pre-Buddhistic conception of salvation, and of the way to be followed by those who would attain thereto.   No more thorough mortification of the flesh has ever been attempted than that achieved by the Indian ascetic anchorite; no later monachism has so nearly succeeded in reducing the human mind to that condition of impassive quasi-somnambulism, which, but for its acknowledged holiness, might run the risk of being confounded with idiocy.

And this salvation, it will be observed, was to be attained through knowledge, and by action based on that knowledge; just as the experimenter, who would obtain a certain physical or chemical result, must have a knowledge of the natural laws involved and the persistent disciplined will adequate to carry out all the various operations required.   The supernatural, in our sense of the term, was entirely excluded.   There was no external power which could affect the sequence of cause and effect which gives rise to karma; none but the will of the subject of the karma which could put an end to it.

Only one rule of conduct could be based upon the remarkable theory of which I have endeavoured to give a reasoned outline.   It was folly to continue

to exist when an overplus of pain was certain;
and the probabilities in favour of the increase of
misery with the prolongation of existence, were
so overwhelming. Slaying the body only made
matters worse; there was nothing for it but to
slay the soul by the voluntary arrest of all its
activities. Property, social ties, family affections,
common companionship, must be abandoned; the
most natural appetites, even that for food, must
be suppressed, or at least minimized; until all
that remained of a man was the impassive,
extenuated, mendicant monk, self-hypnotised
into cataleptic trances, which the deluded mystic
took for foretastes of the final union with
Brahma.

The founder of Buddhism accepted the chief
postulates demanded by his predecessors. But he
was not satisfied with the practical annihilation
involved in merging the individual existence in
the unconditioned—the Atman in Brahma. It
would seem that the admission of the existence of
any substance whatever—even of the tenuity of
that which has neither quality nor energy and of
which no predicate whatever can be asserted—
appeared to him to be a danger and a snare.
Though reduced to a hypostatized negation,
Brahma was not to be trusted; so long as entity
was there, it might conceivably resume the weary
round of evolution, with all its train of immeasurable miseries. Gautama got rid of even that

shade of a shadow of permanent existence by a metaphysical *tour de force* of great interest to the student of philosophy, seeing that it supplies the wanting half of Bishop Berkeley's well-known idealistic argument.

Granting the premises, I am not aware of any escape from Berkeley's conclusion, that the 'substance' of matter is a metaphysical unknown quantity, of the existence of which there is no proof. What Berkeley does not seem to have so clearly perceived is that the non-existence of a substance of mind is equally arguable; and that the result of the impartial applications of his reasonings is the reduction of the All to co-existences and sequences of phenomena, beneath and beyond which there is nothing cognoscible. It is a remarkable indication of the subtlety of Indian speculation that Gautama should have seen deeper than the greatest of modern idealists; though it must be admitted that, if some of Berkeley's reasonings respecting the nature of spirit are pushed home, they reach pretty much the same conclusion.[8]

Accepting the prevalent Brahminical doctrine that the whole cosmos, celestial, terrestrial, and infernal, with its population of gods and other celestial beings, of sentient animals, of Mara and his devils, is incessantly shifting through recurring cycles of production and destruction, in each of which every human being has his transmigratory representative, Gautama proceeded to eliminate substance altogether; and to reduce the cosmos to a mere flow of sensations, emotions, volitions, and thoughts, devoid of any substratum. As, on the surface of a stream of water, we see ripples and whirlpools, which last for a while and then vanish with the causes that gave rise to them, so what seem individual existences are mere temporary associations of phenomena circling round a centre, " like a dog tied to a post." In the whole universe there is nothing permanent, no eternal substance either of mind or of matter. Personality is a metaphysical fancy; and in very truth, not only we, but all things, in the worlds without end of the cosmic phantasmagoria, are such stuff as dreams are made of.

What then becomes of karma ? Karma remains untouched. As the peculiar form of energy we call magnetism may be transmitted from a loadstone to a piece of steel, from the steel to a piece of nickel, as it may be strengthened or weakened by the conditions to which it is subjected while resident in each piece, so it seems to have been conceived that karma might be transmitted from one phenomenal association to another by a sort of induction. However this may be, Gautama doubtless had a better guarantee for the abolition of transmigration, when no wrack of substance, either of Atman or of Brahma, was left behind; when, in short, a man had but to dream that he willed not to dream, to put an end to all dreaming.

This end of life's dream is Nirvana. What Nirvana is the learned do not agree. But, since the best original authorities tell us there is neither desire nor activity, nor any possibility of phenomenal reappearance for the sage who has entered Nirvana, it may be safely said of this acme of Buddhistic philosophy—" the rest is silence."[9]

Thus there is no very great practical disagreement between Gautama and his predecessors with respect to the end of action; but it is otherwise as regards the means to that end. With just insight into human nature, Gautama declared extreme ascetic practices to be useless and indeed harmful. The appetites and the passions are not to be abolished by mere mortification of the body; they must, in addition, be attacked on their own ground and conquered by steady cultivation of the mental habits which oppose them; by universal benevolence; by the return of good for evil; by humility; by abstinence from evil thought; in short, by total renunciation of that self-assertion which is the essence of the cosmic process.

Doubtless, it is to these ethical qualities that Buddhism owes its marvellous success.[10] A system which knows no God in the western sense; which denies a soul to man; which counts the belief in immortality a blunder and the hope of it a sin;

which refuses any efficacy to prayer and sacrifice ; which bids men look to nothing but their own efforts for salvation ; which, in its original purity, knew nothing of vows of obedience, abhorred intolerance, and never sought the aid of the secular arm ; yet spread over a considerable moiety of the Old World with marvellous rapidity, and is still, with whatever base admixture of foreign superstitions, the dominant creed of a large fraction of mankind.

Let us now set our faces westwards, towards Asia Minor and Greece and Italy, to view the rise and progress of another philosophy, apparently independent, but no less pervaded by the conception of evolution.[11]

The sages of Miletus were pronounced evolutionists ; and, however dark may be some of the sayings of Heracleitus of Ephesus, who was probably a contemporary of Gautama, no better expressions of the essence of the modern doctrine of evolution can be found than are presented by some of his pithy aphorisms and striking metaphors.[12]   Indeed, many of my present auditors must have observed that, more than once, I have borrowed from him in the brief exposition of the theory of evolution with which this discourse commenced.

But when the focus of Greek intellectual activity shifted to Athens, the leading minds concentrated their attention upon ethical problems. Forsaking the study of the macrocosm for that of the microcosm, they lost the key to the thought of the great Ephesian, which, I imagine, is more intelligible to us than it was to Socrates, or to Plato. Socrates, more especially, set the fashion of a kind of inverse agnosticism, by teaching that the problems of physics lie beyond the reach of the human intellect; that the attempt to solve them is essentially vain; that the one worthy object of investigation is the problem of ethical life; and his example was followed by the Cynics and the later Stoics. Even the comprehensive knowledge and the penetrating intellect of Aristotle failed to suggest to him that in holding the eternity of the world, within its present range of mutation, he was making a retrogressive step. The scientific heritage of Heracleitus passed into the hands neither of Plato nor of Aristotle, but into those of Democritus. But the world was not yet ready to receive the great conceptions of the philosopher of Abdera. It was reserved for the Stoics to return to the track marked out by the earlier philosophers; and, professing themselves disciples of Heracleitus, to develop the idea of evolution systematically. In doing this, they not only omitted some characteristic features of their master's teaching, but they made additions altogether foreign to it. One of the most influential of these importations was the transcendental theism which had come into vogue. The restless, fiery energy, operating according to law, out of which all things emerge and into which they return, in the endless successive cycles of the great year; which creates and destroys worlds as a wanton child builds up, and anon levels, sand castles on the seashore; was metamorphosed into a material world-soul and decked out with all the attributes of ideal Divinity; not merely with infinite power and transcendent wisdom, but with absolute goodness.

The consequences of this step were momentous. For if the cosmos is the effect of an immanent, omnipotent, and infinitely beneficent cause, the existence in it of real evil, still less of necessarily inherent evil, is plainly inadmissible.[13] Yet the universal experience of mankind testified then, as now, that, whether we look within us or without us, evil stares us in the face on all sides; that if anything is real, pain and sorrow and wrong are realities.

It would be a new thing in history if *à priori* philosophers were daunted by the factious opposition of experience; and the Stoics were the last men to allow themselves to be beaten by mere facts. 'Give me a doctrine and I will find the reasons for it,' said Chrysippus. So they perfected, if they did not invent, that ingenious and plausible form of pleading, the Theodicy; for the purpose of showing firstly, that there is no such thing as evil; secondly, that if there is, it is the necessary correlate of good; and, moreover, that it is either due to our own fault, or inflicted for our benefit. Theodicies have been very popular in their time, and I believe that a numerous, though somewhat dwarfed, progeny of them still survives. So far as I know, they are all variations of the theme set forth in those famous six lines of the "Essay on Man," in which Pope sums up Bolingbroke's reminiscences of stoical and other speculations of this kind—

"All nature is but art, unknown to thee;
　All chance, direction which thou canst not see;
　All discord, harmony not understood;
　All partial evil, universal good;
　And spite of pride, in erring reason's spite,
　One truth is clear: whatever is is right."

Yet, surely, if there are few more important truths than those enunciated in the first triad, the second is open to very grave objections. That there is a 'soul of good in things evil' is unquestionable; nor will any wise man deny the disciplinary value of pain and sorrow. But these considerations do not help us to see why the immense multitude of irresponsible sentient beings, which cannot profit by such discipline, should suffer; nor why, among the endless possibilities open to omnipotence—that of sinless, happy existence among the rest—the actuality in which sin and misery abound should be that selected.

Surely it is mere cheap rhetoric to call arguments which have never yet been answered by even the meekest and the least rational of Optimists, suggestions of the pride of reason. As to the concluding aphorism, its fittest place would be as an inscription in letters of mud over the portal of some ' stye of Epicurus ' ; [14] for that is where the logical application of it to practice would land men, with every aspiration stifled and every effort paralyzed. Why try to set right what is right already? Why strive to improve the best of all possible worlds ? Let us eat and drink, for as to-day all is right, so to-morrow all will be.

But the attempt of the Stoics to blind themselves to the reality of evil, as a necessary concomitant of the cosmic process, had less success than that of the Indian philosophers to exclude the reality of good from their purview. Unfortunately, it is much easier to shut one's eyes to good than to evil. Pain and sorrow knock at our doors more loudly than pleasure and happiness ; and the prints of their heavy footsteps are less easily effaced. Before the grim realities of practical life the pleasant fictions of optimism vanished. If this were the best of all possible worlds, it nevertheless proved itself a very inconvenient habitation for the ideal sage.

The stoical summary of the whole duty of man, ' Live according to nature,' would seem to imply that the cosmic process is an exemplar for human conduct. Ethics would thus become applied Natural History. In fact, a confused employment of the maxim, in this sense, has done immeasurable mischief in later times. It has furnished an axiomatic foundation for the philosophy of philosophasters and for the moralizing of sentimentalists. But the Stoics were, at bottom, not merely noble, but sane, men; and if we look closely into what they really meant by this ill-used phrase, it will be found to present no justification for the mischievous conclusions that have been deduced from it.

In the language of the Stoa, 'Nature' was a word of many meanings. There was the 'Nature' of the cosmos and the 'Nature' of man. In the latter, the animal 'nature,' which man shares with a moiety of the living part of the cosmos, was distinguished from a higher 'nature.' Even in this higher nature there were grades of rank. The logical faculty is an instrument which may be turned to account for any purpose. The passions and the emotions are so closely tied to the lower nature that they may be considered to be pathological, rather than normal, phenomena. The one supreme, hegemonic, faculty, which constitutes the essential 'nature' of man, is most nearly represented by that which, in the language of a later philosophy, has been called the pure reason. It is this 'nature' which holds up the ideal of the supreme good and demands absolute submission of

the will to its behests. It is this which commands
all men to love one another, to return good for evil, to
regard one another as fellow-citizens of one great
state. Indeed, seeing that the progress towards
perfection of a civilized state, or polity, depends
on the obedience of its members to these commands, the Stoics sometimes termed the pure
reason the 'political' nature. Unfortunately,
the sense of the adjective has undergone so much
modification, that the application of it to that
which commands the sacrifice of self to the
common good would now sound almost grotesque.[15]

But what part is played by the theory of evolution in this view of ethics ? So far as I can
discern, the ethical system of the Stoics, which is
essentially intuitive, and reverences the categorical
imperative as strongly as that of any later
moralists, might have been just what it was if they
had held any other theory ; whether that of special
creation, on the one side, or that of the eternal
existence of the present order, on the other.[16] To
the Stoic, the cosmos had no importance for the
conscience, except in so far as he chose to think
it a pedagogue to virtue. The pertinacious optimism of our philosophers hid from them the actual
state of the case. It prevented them from seeing
that cosmic nature is no school of virtue, but the
headquarters of the enemy of ethical nature.
The logic of facts was necessary to convince them

that the cosmos works through the lower nature of man, not for righteousness, but against it. And it finally drove them to confess that the existence of their ideal " wise man " was incompatible with the nature of things ; that even a passable approximation to that ideal was to be attained only at the cost of renunciation of the world and mortification, not merely of the flesh, but of all human affections. The state of perfection was that 'apatheia '[17] in which desire, though it may still be felt, is powerless to move the will, reduced to the sole function of executing the commands of pure reason. Even this residuum of activity was to be regarded as a temporary loan, as an efflux of the divine world-pervading spirit, chafing at its imprisonment in the flesh, until such time as death enabled it to return to its source in the all-pervading logos.

I find it difficult to discover any very great difference between Apatheia and Nirvana, except that stoical speculation agrees with pre-Buddhistic philosophy, rather than with the teachings of Gautama, in so far as it postulates a permanent substance equivalent to ' Brahma ' and ' Atman '; and that, in stoical practice, the adoption of the life of the mendicant cynic was held to be more a counsel of perfection than an indispensable condition of the higher life.

Thus the extremes touch. Greek thought and

Indian thought set out from ground common to both, diverge widely, develop under very different physical and moral conditions, and finally converge to practically the same end.

The Vedas and the Homeric epos set before us a world of rich and vigorous life, full of joyous fighting men

> That ever with a frolic welcome took
> The thunder and the sunshine . . . .

and who were ready to brave the very Gods themselves when their blood was up. A few centuries pass away, and under the influence of civilization the descendants of these men are ' sicklied o'er with the pale cast of thought'—frank pessimists, or, at best, make-believe optimists. The courage of the warlike stock may be as hardly tried as before, perhaps more hardly, but the enemy is self. The hero has become a monk. The man of action is replaced by the quietist, whose highest aspiration is to be the passive instrument of the divine Reason. By the Tiber, as by the Ganges, ethical man admits that the cosmos is too strong for him; and, destroying every bond which ties him to it by ascetic discipline, he seeks salvation in absolute renunciation.[18]

Modern thought is making a fresh start from the base whence Indian and Greek philosophy set out; and, the human mind being very much what

it was six-and-twenty centuries ago, there is no
ground for wonder if it presents indications of a
tendency to move along the old lines to the same
results.

We are more than sufficiently familiar with
modern pessimism, at least as a speculation; for I
cannot call to mind that any of its present votaries
have sealed their faith by assuming the rags and
the bowl of the mendicant Bhikku, or the cloak
and the wallet of the Cynic. The obstacles placed
in the way of sturdy vagrancy by an unphiloso-
phical police have, perhaps, proved too formidable
for philosophical consistency. We also know
modern speculative optimism, with its perfectibility
of the species, reign of peace, and lion and lamb
transformation scenes; but one does not hear so
much of it as one did forty years ago; indeed, I
imagine it is to be met with more commonly at
the tables of the healthy and wealthy, than in the
congregations of the wise. The majority of us, I
apprehend, profess neither pessimism nor optimism.
We hold that the world is neither so good, nor so
bad, as it conceivably might be; and, as most of
us have reason, now and again, to discover that it
can be. Those who have failed to experience the
joys that make life worth living are, probably, in
as small a minority as those who have never
known the griefs that rob existence of its savour
and turn its richest fruits into mere dust and
ashes.

Further, I think I do not err in assuming that, however diverse their views on philosophical and religious matters, most men are agreed that the proportion of good and evil in life may be very sensibly affected by human action. I never heard anybody doubt that the evil may be thus increased, or diminished; and it would seem to follow that good must be similarly susceptible of addition or subtraction. Finally, to my knowledge, nobody professes to doubt that, so far forth as we possess a power of bettering things, it is our paramount duty to use it and to train all our intellect and energy to this supreme service of our kind.

Hence the pressing interest of the question, to what extent modern progress in natural knowledge, and, more especially, the general outcome of that progress in the doctrine of evolution, is competent to help us in the great work of helping one another?

The propounders of what are called the " ethics of evolution," when the ' evolution of ethics ' would usually better express the object of their speculations, adduce a number of more or less interesting facts and more or less sound arguments, in favour of the origin of the moral sentiments, in the same way as other natural phenomena, by a process of evolution. I have little doubt, for my own part, that they are on the right track; but as the immoral sentiments have no less been evolved, there is, so far, as much natural sanction for the

one as the other. The thief and the murderer
follow nature just as much as the philanthropist.
Cosmic evolution may teach us how the good and
the evil tendencies of man may have come about;
but, in itself, it is incompetent to furnish any
better reason why what we call good is preferable
to what we call evil than we had before. Some
day, I doubt not, we shall arrive at an understanding of the evolution of the æsthetic faculty; but
all the understanding in the world will neither
increase nor diminish the force of the intuition
that this is beautiful and that is ugly.

There is another fallacy which appears to me to
pervade the so-called " ethics of evolution." It is
the notion that because, on the whole, animals
and plants have advanced in perfection of organization by means of the struggle for existence and
the consequent 'survival of the fittest'; therefore
men in society, men as ethical beings, must look
to the same process to help them towards perfection. I suspect that this fallacy has arisen out
of the unfortunate ambiguity of the phrase 'survival of the fittest.' 'Fittest' has a connotation of
'best'; and about 'best' there hangs a moral
flavour. In cosmic nature; however, what is
' fittest ' depends upon the conditions. Long since,[19]
I ventured to point out that if our hemisphere
were to cool again, the survival of the fittest might
bring about, in the vegetable kingdom, a population of more and more stunted and humbler and

humbler organisms, until the 'fittest' that survived might be nothing but lichens, diatoms, and
such microscopic organisms as those which give
red snow its colour; while, if it became hotter, the
pleasant valleys of the Thames and Isis might be
uninhabitable by any animated beings save those
that flourish in a tropical jungle. They, as the
fittest, the best adapted to the changed conditions,
would survive.

Men in society are undoubtedly subject to the
cosmic process. As among other animals, multiplication goes on without cessation, and involves
severe competition for the means of support. The
struggle for existence tends to eliminate those less
fitted to adapt themselves to the circumstances
of their existence. The strongest, the most self-
assertive, tend to tread down the weaker. But
the influence of the cosmic process on the evolution of society is the greater the more rudimentary
its civilization. Social progress means a checking
of the cosmic process at every step and the substitution for it of another, which may be called
the ethical process; the end of which is not the
survival of those who may happen to be the
fittest, in respect of the whole of the conditions
which obtain, but of those who are ethically the
best.[20]

As I have already urged, the practice of that
which is ethically best—what we call goodness or
virtue—involves a course of conduct which, in all

respects, is opposed to that which leads to success in the cosmic struggle for existence. In place of ruthless self-assertion it demands self-restraint; in place of thrusting aside, or treading down, all competitors, it requires that the individual shall not merely respect, but shall help his fellows; its influence is directed, not so much to the survival of the fittest, as to the fitting of as many as possible to survive. It repudiates the gladiatorial theory of existence. It demands that each man who enters into the enjoyment of the advantages of a polity shall be mindful of his debt to those who have laboriously constructed it; and shall take heed that no act of his weakens the fabric in which he has been permitted to live. Laws and moral precepts are directed to the end of curbing the cosmic process and reminding the individual of his duty to the community, to the protection and influence of which he owes, if not existence itself, at least the life of something better than a brutal savage.

It is from neglect of these plain considerations that the fanatical individualism [21] of our time attempts to apply the analogy of cosmic nature to society. Once more we have a misapplication of the stoical injunction to follow nature; the duties of the individual to the state are forgotten, and his tendencies to self-assertion are dignified by the name of rights. It is seriously debated whether the members of a community are justified in

using their combined strength to constrain one of
their number to contribute his share to the maintenance of it; or even to prevent him from doing
his best to destroy it. The struggle for existence,
which has done such admirable work in cosmic
nature, must, it appears, be equally beneficent in
the ethical sphere. Yet if that which I have insisted upon is true; if the cosmic process has no
sort of relation to moral ends; if the imitation of
it by man is inconsistent with the first principles
of ethics; what becomes of this surprising theory?

Let us understand, once for all, that the ethical
progress of society depends, not on imitating the
cosmic process, still less in running away from it,
but in combating it. It may seem an audacious
proposal thus to pit the microcosm against the
macrocosm and to set man to subdue nature to his
higher ends; but I venture to think that the
great intellectual difference between the ancient
times with which we have been occupied and our
day, lies in the solid foundation we have acquired
for the hope that such an enterprise may meet
with a certain measure of success.

The history of civilization details the steps by
which men have succeeded in building up an
artificial world within the cosmos. Fragile reed
as he may be, man, as Pascal says, is a thinking
reed: [22] there lies within him a fund of energy,
operating intelligently and so far akin to that
which pervades the universe, that it is competent

to influence and modify the cosmic process. In virtue of his intelligence, the dwarf bends the Titan to his will. In every family, in every polity that has been established, the cosmic process in man has been restrained and otherwise modified by law and custom; in surrounding nature, it has been similarly influenced by the art of the shepherd, the agriculturist, the artisan. As civilization has advanced, so has the extent of this interference increased; until the organized and highly developed sciences and arts of the present day have endowed man with a command over the course of non-human nature greater than that once attributed to the magicians. The most impressive, I might say startling, of these changes have been brought about in the course of the last two centuries; while a right comprehension of the process of life and of the means of influencing its manifestations is only just dawning upon us. We do not yet see our way beyond generalities; and we are befogged by the obtrusion of false analogies and crude anticipations. But Astronomy, Physics, Chemistry, have all had to pass through similar phases, before they reached the stage at which their influence became an important factor in human affairs. Physiology, Psychology, Ethics, Political Science, must submit to the same ordeal. Yet it seems to me irrational to doubt that, at no distant period, they will work as great a revolution in the sphere of practice.

The theory of evolution encourages no millennial anticipations. If, for millions of years, our globe has taken the upward road, yet, some time, the summit will be reached and the downward route will be commenced. The most daring imagination will hardly venture upon the suggestion that the power and the intelligence of man can ever arrest the procession of the great year.

Moreover, the cosmic nature born with us and, to a large extent, necessary for our maintenance, is the outcome of millions of years of severe training, and it would be folly to imagine that a few centuries will suffice to subdue its masterfulness to purely ethical ends. Ethical nature may count upon having to reckon with a tenacious and powerful enemy as long as the world lasts. But, on the other hand, I see no limit to the extent to which intelligence and will, guided by sound principles of investigation, and organized in common effort, may modify the conditions of existence, for a period longer than that now covered by history. And much may be done to change the nature of man himself.[23] The intelligence which has converted the brother of the wolf into the faithful guardian of the flock ought to be able to do something towards curbing the instincts of savagery in civilized men.

But if we may permit ourselves a larger hope of abatement of the essential evil of the world than was possible to those who, in the infancy of exact knowledge, faced the problem of existence more than a score of centuries ago, I deem it an essential condition of the realization of that hope that we should cast aside the notion that the escape from pain and sorrow is the proper object of life.

We have long since emerged from the heroic childhood of our race, when good and evil could be met with the same 'frolic welcome'; the attempts to escape from evil, whether Indian or Greek, have ended in flight from the battle-field; it remains to us to throw aside the youthful over-confidence and the no less youthful discouragement of nonage. We are grown men, and must play the man

> strong in will
> To strive, to seek, to find, and not to yield,

cherishing the good that falls in our way, and bearing the evil, in and around us, with stout hearts set on diminishing it. So far, we all may strive in one faith towards one hope:

> It may be that the gulfs will wash us down,
> It may be we shall touch the Happy Isles,
>
> . . . . but something ere the end,
> Some work of noble note may yet be done. ([24])

NOTES

Note 1 (p. 49).

I HAVE been careful to speak of the "appearance" of cyclical evolution presented by living things; for, on critical examination, it will be found that the course of vegetable and of animal life is not exactly represented by the figure of a cycle which returns into itself. What actually happens, in all but the lowest organisms, is that one part of the growing germ (A) gives rise to tissues and organs; while another part (B) remains in its primitive condition, or is but slightly modified. The moiety A becomes the body of the adult and, sooner or later, perishes, while portions of the moiety B are detached and, as offspring, continue the life of the species. Thus, if we trace back an organism along the direct line of descent from its remotest ancestor, B, as a whole, has never suffered death; portions of it, only, have been cast off and died in each individual offspring.

Everybody is familiar with the way in which the "suckers" of a strawberry plant behave. A thin cylinder of living tissue keeps on growing at its free end, until it attains a considerable length. At successive intervals, it develops buds which grow into strawberry plants ; and these become independent by the death of the parts of the sucker which connect them. The rest of the sucker, however, may go on living and growing indefinitely, and, circumstances remaining favourable, there is no obvious reason why it should ever die. The living substance *B*, in a manner, answers to the sucker. If we could restore the continuity which was once possessed by the portions of *B*, contained in all the individuals of a direct line of descent, they would form a sucker, or *stolon*, on which these individuals would be strung, and which would never have wholly died.

A species remains unchanged so long as the potentiality of development resident in *B* remains unaltered ; so long, *e.g.*, as the buds of the strawberry sucker tend to become typical strawberry plants. In the case of the progressive evolution of a species, the developmental potentiality of *B* becomes of a higher and higher order. In retrogressive evolution, the contrary would be the case. The phenomena of atavism seem to show that retrogressive evolution, that is, the return of a species to one or other of its earlier forms, is a possibility to be reckoned with. The simplification of structure, which is so common in the parasitic members of a group, however, does not properly come under this head. The worm-like, limbless *Lernœa* has no resemblance to any of the stages of development of the many-limbed active animals of the group to which it belongs.

Note 2 (p. 49).

Heracleitus says, Ποταμῷ γὰρ οὐκ ἔστι δὶς ἐμβῆναι τῷ αὐτῷ; but, to be strictly accurate, the river remains, though the water of which it is composed changes—just as a man retains his identity though the whole substance of his body is constantly shifting. This is put very well by Seneca (Ep. lvii. i. 20, Ed. Ruhkopf): "Corpora nostra rapiuntur fluminum more, quidquid vides currit cum tempore; nihil ex his quæ videmus manet. Ego ipse dum loquor mutari ista, mutatus sum. Hoc est quod ait Heraclitus ' In idem flumen bis non descendimus.' Manet idem fluminis nomen, aqua transmissa est. Hoc in amne manifestius est quam in homine, sed nos quoque non minus velox cursus prætervehit."

Note 3 (p. 55).

"Multa bona nostra nobis nocent, timoris enim tormentum memoria reducit, providentia anticipat. Nemo tantum præsentibus miser est." (Seneca, Ed. v. 7.)

Among the many wise and weighty aphorisms of the Roman Bacon, few sound the realities of life more deeply than "Multa bona nostra nobis nocent." If there is a soul of good in things evil, it is at least equally true that there is a soul of evil in things good : for things, like men, have "les défauts de leurs qualités." It is one of the last lessons one learns from experience, but not the least important, that a

heavy tax is levied upon all forms of success; and
that failure is one of the commonest disguises
assumed by blessings.

Note 4 (p. 60).

" There is within the body of every man a soul
which, at the death of the body, flies away from. it
like a bird out of a cage, and enters upon a new
life . . . either in one of the heavens or one of the
hells or on this earth. The only exception is the
rare case of a man having in this life acquired a
true knowledge of God. According to the pre-
Buddhistic theory, the soul of such a man goes along
the path of the Gods to God, and, being united with
Him, enters upon an immortal life in which his
individuality is not extinguished. In the latter theory,
his soul is directly absorbed into the Great Soul, is
lost in it, and has no longer any independent existence.
The souls of all other men enter, after the death of
the body, upon a new existence in one or other of
the many different modes of being. If in heaven or
hell, the soul itself becomes a god or demon without
entering a body; all superhuman beings, save the
great gods, being looked upon as not eternal, but
merely temporary creatures. If the soul returns to
earth it may or may not enter a new body; and this
either of a human being, an animal, a plant, or even
a material object. For all these are possessed of
souls, and there is no essential difference between
these souls and the souls of men—all being alike
mere sparks of the Great Spirit, who is the only real

existence." (Rhys Davids, *Hibbert Lectures*, 1881, p. 83.)

For what I have said about Indian Philosophy, I am particularly indebted to the luminous exposition of primitive Buddhism and its relations to earlier Hindu thought, which is given by Prof. Rhys Davids in his remarkable *Hibbert Lectures* for 1881, and *Buddhism* (1890). The only apology I can offer for the freedom with which I have borrowed from him in these notes, is my desire to leave no doubt as to my indebtedness. I have also found Dr. Oldenberg's *Buddha* (Ed. 2, 1890) very helpful. The origin of the theory of transmigration stated in the above extract is an unsolved problem. That it differs widely from the Egyptian metempsychosis is clear. In fact, since men usually people the other world with phantoms of this, the Egyptian doctrine would seem to presuppose the Indian as a more archaic belief.

Prof. Rhys Davids has fully insisted upon the ethical importance of the transmigration theory. " One of the latest speculations now being put forward among ourselves would seek to explain each man's character, and even his outward condition in life, by the character he inherited from his ancestors, a character gradually formed during a practically endless series of past existences, modified only by the conditions into which he was born, those very conditions being also, in like manner, the last result of a practically endless series of past causes. Gotama's speculation might be stated in the same words. But it attempted also to explain, in a way different from

that which would be adopted by the exponents of
the modern theory, that strange problem which it
is also the motive of the wonderful drama of the
book of Job to explain—the fact that the actual
distribution here of good fortune, or misery, is
entirely independent of the moral qualities which
men call good or bad. We cannot wonder that a
teacher, whose whole system was so essentially an
ethical reformation, should have felt it incumbent
upon him to seek an explanation of this apparent
injustice. And all the more so, since the belief he
had inherited, the theory of the transmigration of
souls, had provided a solution perfectly sufficient to
any one who could accept that belief." (*Hibbert
Lectures*, p. 93.) I should venture to suggest the
substitution of 'largely' for 'entirely' in the foregoing passage. Whether a ship makes a good or a
bad voyage is largely independent of the conduct of
the captain, but it is largely affected by that conduct. Though powerless before a hurricane he may
weather many a bad gale.

Note 5 (p. 61).

The outward condition of the soul is, in each new
birth, determined by its actions in a previous birth;
but by each action in succession, and not by the
balance struck after the evil has been reckoned off
against the good. A good man who has once uttered
a slander may spend a hundred thousand years as a
god, in consequence of his goodness, and when the
power of his good actions is exhausted, may be born

as a dumb man on account of his transgression ; and a robber who has once done an act of mercy, may come to life in a king's body as the result of his virtue, and then suffer torments for ages in hell or as a ghost without a body, or be re-born many times as a slave or an outcast, in consequence of his evil life.

" There is no escape, according to this theory, from the result of any act; though it is only the consequences of its own acts that each soul has to endure. The force has been set in motion by itself and can never stop ; and its effect can never be foretold. If evil, it can never be modified or prevented, for it depends on a cause already completed, that is now for ever beyond the soul's control. There is even no continuing consciousness, no memory of the past that could guide the soul to any knowledge of its fate. The only advantage open to it is to add in this life to the sum of its good actions, that it may bear fruit with the rest. And even this can only happen in some future life under essentially the same conditions as the present one : subject, like the present one, to old age, decay, and death ; and affording opportunity, like the present one, for the commission of errors, ignorances, or sins, which in their turn must inevitably produce their due effect of sickness, disability, or woe. Thus is the soul tossed about from life to life, from billow to billow in the great ocean of transmigration. And there is no escape save for the very few, who, during their birth as men, attain to a right knowledge of the Great Spirit : and thus enter into immortality, or, as the later philosophers taught, are absorbed into the

Divine Essence." (Rhys Davids, *Hibbert Lectures*, pp. 85, 86.)

The state after death thus imagined by the Hindu philosophers has a certain analogy to the purgatory of the Roman Church; except that escape from it is dependent, not on a divine decree modified, it may be, by sacerdotal or saintly intercession, but by the acts of the individual himself; and that while ultimate emergence into heavenly bliss of the good, or well-prayed for, Catholic is professedly assured, the chances in favour of the attainment of absorption, or of Nirvana, by any individual Hindu are extremely small.

Note 6 (p. 62).

" That part of the then prevalent transmigration theory which could not be proved false seemed to meet a deeply felt necessity, seemed to supply a moral cause which would explain the unequal distribution here of happinesss or woe, so utterly inconsistent with the present characters of men." Gautama "still therefore talked of men's previous existence, but by no means in the way that he is generally represented to have done." What he taught was "the transmigration of character." He held that after the death of any being, whether human or not, there survived nothing at all but that being's 'Karma,' the result, that is, of its mental and bodily actions. Every individual, whether human or divine, was the last inheritor and the last result of the Karma of a long series of past individuals — a series so long that its beginning is beyond the reach of calculation, and its end will be coincident with the destruction of the world." (Rhys Davids, *Hibbert Lectures*, p. 92.)

In the theory of evolution, the tendency of a germ to develop according to a certain specific type, *e.g.* of the kidney bean seed to grow into a plant having all the characters of *Phaseolus vulgaris*, is its 'Karma.' It is the " last inheritor and the last result " of all the conditions that have affected a line of ancestry which goes back for many millions of years to the time when life first appeared on the earth. The moiety B of the substance of the bean plant (see *Note* 1) is the last link in a once continuous chain extending from the primitive living substance : and the characters of the successive species to which it has given rise are the manifestations of its gradually modified Karma. As Prof. Rhys Davids aptly says, the snowdrop " is a snowdrop and not an oak, and just that kind of snowdrop, because it is the outcome of the Karma of an endless series of past existences." (*Hibbert Lectures*, p. 114.)

Note 7 (p. 64).

" It is interesting to notice that the very point which is the weakness of the theory—the supposed concentration of the effect of the Karma in one new being—presented itself to the early Buddhists them-selves as a difficulty. They avoided it, partly by explaining that it was a particular thirst in the creature dying (a craving, Tanhā, which plays otherwise a great part in the Buddhist theory) which actually caused the birth of the new individual who was to inherit the Karma of the former one. But, how this took place, how the craving desire produced this effect, was acknowledged to be a mystery patent only to a Buddha." (Rhys Davids, *Hibbert Lectures*, p. 95.)

Among the many parallelisms of Stoicism and Buddhism, it is curious to find one for this Tanhā, 'thirst,' or 'craving desire' for life. Seneca writes (Epist. lxxvi. 18): "Si enim ullum aliud est bonum quam honestum, sequetur nos *aviditas vitæ* aviditas rerum vitam instruentium: quod est intolerabile infinitum, vagum."

Note 8 (p. 66).

"The distinguishing characteristic of Buddhism was that it started a new line, that it looked upon the deepest questions men have to solve from an entirely different standpoint. It swept away from the field of its vision the whole of the great soul-theory which had hitherto so completely filled and dominated the minds of the superstitious and the thoughtful alike. For the first time in the history of the world, it proclaimed a salvation which each man could gain for himself and by himself, in this world, during this life, without any the least reference to God, or to Gods, either great or small. Like the Upanishads, it placed the first importance on know-ledge; but it was no longer a knowledge of God, it was a clear perception of the real nature, as they supposed it to be, of men and things. And it added to the necessity of knowledge, the necessity of purity, of courtesy, of uprightness, of peace and of a universal love far reaching, grown great and beyond measure." (Rhys Davids, *Hibbert Lectures*, p. 29.)

The contemporary Greek philosophy takes an analogous direction. According to Heracleitus, the universe was made neither by Gods nor men; but, from all eternity, has been, and to all eternity, will be, immortal fire, glowing and fading in due measure. (Mullach, *Heracliti Fragmenta*, 27.) And the part assigned by his successors, the Stoics, to the knowledge and the volition of the ' wise man' made their Divinity (for logical thinkers) a subject for compliments, rather than a power to be reckoned with. In Hindu speculation the ' Arahat,' still more the ' Buddha,' becomes the superior of Brahma; the stoical ' wise man' is, at least, the equal of Zeus.

Berkeley affirms over and over again that no idea can be formed of a soul or spirit—" If any man shall doubt of the truth of what is here delivered, let him but reflect and try if he can form any idea of power or active being; and whether he hath ideas of two principal powers marked by the names of *will* and *understanding* distinct from each other, as well as from a third idea of substance or being in general, with a relative notion of its supporting or being the subject of the aforesaid power, which is signified by the name *soul* or *spirit*. This is what some hold: but, so far as I can see, the words *will, soul, spirit,* do not stand for different ideas or, in truth, for any idea at all, but for something which is very different from ideas, and which, being an agent, cannot be like unto or represented by any idea whatever [though it must be owned at the same time, that we have some notion of soul, spirit, and the operations of the mind, such as willing, loving, hating, inasmuch as we know or understand the meaning of these words "]. (*The Principles of Human Knowledge*, lxxvi. See also §§ lxxxix., cxxxv., cxlv.)

It is open to discussion, I think, whether it is possible to have 'some notion' of that of which we can form no 'idea.'

Berkeley attaches several predicates to the "perceiving active being mind, spirit, soul or myself" (Parts I. II.) It is said, for example, to be "indivisible, incorporeal, unextended, and incorruptible." The predicate indivisible, though negative in form, has highly positive consequences. For, if 'perceiving active being' is strictly indivisible, man's soul must be one with the Divine spirit: which is good Hindu or Stoical doctrine, but hardly orthodox Christian philosophy. If, on the other hand, the 'substance' of active perceiving 'being' is actually divided into the one Divine and innumerable human entities, how can the predicate 'indivisible' be rigorously applicable to it?

Taking the words cited, as they stand, they amount to the denial of the possibility of any knowledge of substance. 'Matter' having been resolved into mere affections of 'spirit,' 'spirit' melts away into an admittedly inconceivable and unknowable hypostasis of thought and power—consequently the existence of anything in the universe beyond a flow of phenomena is a purely hypothetical assumption. Indeed a pyrrhonist might raise the objection that if 'esse' is 'percipi' spirit itself can have no existence except as a perception, hypostatized into a 'self,' or as a perception of some other spirit. In the former case, objective reality vanishes; in the latter, there would seem to be the need of an infinite series of spirits each perceiving the others.

It is curious to observe how very closely the phraseology of Berkeley sometimes approaches that of the Stoics: thus (cxlviii.) " It seems to be a *general pretence of the unthinking* herd that *they cannot see God*......But, alas, we need only open our eyes to see the Sovereign Lord of all things with a more full and clear view, than we do any of our fellow-creatures......we do at all times and in all places perceive manifest tokens of the Divinity: everything we see, hear, feel, or any wise perceive by sense, being a sign or effect of the power of God " ......cxlix. " It is therefore plain, that *nothing can be more evident* to any one that is capable of the least reflection, *than the existence of God*, or a spirit who is intimately present to our minds, producing in them all that variety of ideas or sensations which continually affect us, on whom we have an absolute and entire dependence, in short, *in whom we live and move and have our being.*" cl. [But you will say hath Nature no share in the production of natural things, and must they be all ascribed to the immediate and sole operation of God?......if by *Nature* is meant some

being distinct from God, as well as from the laws of
nature and things perceived by sense, I must confess
that word is to me an empty sound, without any
intelligible meaning annexed to it.] Nature in this
acceptation is a vain *Chimæra* introduced by those
heathens, who had not just notions of the omni-
presence and infinite perfection of God."

Compare Seneca (*De Beneficiis*, iv. 7) :

"Natura, inquit, hæc mihi præstat. Non intelligis
te, quum hoc dicis, mutare Nomen Deo? Quid enim
est aliud Natura quam Deus, et divina ratio, toti
mundo et partibus ejus inserta? Quoties voles tibi
licet aliter hunc auctorem rerum nostrarum com-
pellare, et Jovem illum optimum et maximum rite
dices, et tonantem, et statorem : qui non, ut historici
tradiderunt, ex eo quod post votum susceptum acies
Romanorum fugientum stetit, sed quod stant beneficio
ejus omnia, stator, stabilitorque est : hunc eundem et
fatum si dixeris, non mentieris, nam quum fatum
nihil aliud est, quam series implexa causarum, ille est
prima omnium causa, ea qua cæteræ pendent." It
would appear, therefore, that the good Bishop is
somewhat hard upon the 'heathen,' of whose words
his own might be a paraphrase.

There is yet another direction in which Berkeley's
philosophy, I will not say agrees with Gautama's, but
at any rate helps to make a fundamental dogma of
Buddhism intelligible.

"I find I can excite ideas in my mind at pleasure,
and vary and shift the scene as often as I think fit.
It is no more than willing, and straightway this or
that idea arises in my fancy : and by the same power,

it is obliterated, and makes way for another. This making and unmaking of ideas doth very properly denominate the mind active. This much is certain and grounded on experience. . . ." (*Principles*, xxviii.)

A good many of us, I fancy, have reason to think that experience tells them very much the contrary; and are painfully familiar with the obsession of the mind by ideas which cannot be obliterated by any effort of the will and steadily refuse to make way for others. But what I desire to point out is that if Gautama was equally confident that he could 'make and unmake' ideas—then, since he had resolved self into a group of ideal phantoms—the possibility of abolishing self by volition naturally followed.

Note 9 (p. 68).

According to Buddhism, the relation of one life to the next is merely that borne by the flame of one lamp to the flame of another lamp which is set alight by it. To the 'Arahat' or adept "no outward form, no compound thing, no creature, no creator, no existence of any kind, must appear to be other than a temporary collocation of its component parts, fated inevitably to be dissolved."—(Rhys Davids, *Hibbert Lectures*, p. 211.)

The self is nothing but a group of phenomena held together by the desire of life; when that desire shall have ceased, "the Karma of that particular chain of lives will cease to influence any longer any distinct individual, and there will be no more birth; for birth, decay, and death, grief, lamentation, and despair will have come, so far as regards that chain of lives, for ever to an end."

The state of mind of the Arahat in which the desire of life has ceased is Nirvana. Dr. Oldenberg has very acutely and patiently considered the various interpretations which have been attached to ' Nirvana' in the work to which I have referred (pp. 285 *et seq.*). The result of his and other discussions of the question may I think be briefly stated thus :

1. Logical deduction from the predicates attached to the term 'Nirvana' strips it of all reality, conceivability, or perceivability, whether by Gods or men. For all practical purposes, therefore, it comes to exactly the same thing as annihilation.

2. But it is not annihilation in the ordinary sense, inasmuch as it could take place in the living Arahat or Buddha.

3. And, since, for the faithful Buddhist, that which was abolished in the Arahat was the possibility of further pain, sorrow, or sin ; and that which was attained was perfect peace ; his mind directed itself exclusively to this joyful consummation, and personified the negation of all conceivable existence and of all pain into a positive bliss. This was all the more easy, as Gautama refused to give any dogmatic definition of Nirvana. There is something analogous in the way in which people commonly talk of the ' happy release' of a man who has been long suffering from mortal disease. According to their own views, it must always be extremely doubtful whether the man will be any happier after the ' release' than

before. But they do not choose to look at the matter in this light.

The popular notion that, with practical, if not metaphysical, annihilation in view, Buddhism must needs be a sad and gloomy faith seems to be inconsistent with fact; on the contrary, the prospect of Nirvana fills the true believer, not merely with cheerfulness, but with an ecstatic desire to reach it.

Note 10 (p. 68).

The influence of the picture of the personal quali·ties of Gautama, afforded by the legendary anecdotes which rapidly grew into a biography of the Buddha; and by the birth stories, which coalesced with the current folk-lore, and were intelligible to all the world, doubtless played a great part. Further, although Gautama appears not to have meddled with the caste system, he refused to recognize any distinction, save that of perfection in the way of salvation, among his followers; and by such teaching, no less than by the inculcation of love and benevolence to all sentient beings, he practically levelled every social, political, and racial barrier. A third important condition was the organization of the Buddhists into monastic communities for the stricter professors, while the laity were permitted a wide indulgence in practice and were allowed to hope for accommodation in some of the temporary abodes of bliss. With a few hundred thousand years of immediate paradise in sight, the average man could be content to shut his eyes to what might follow.

Note 11 (p. 69).

In ancient times it was the fashion, even among the Greeks themselves, to derive all Greek wisdom from Eastern sources; not long ago it was as generally denied that Greek philosophy had any connection with Oriental speculation; it seems probable, however, that the truth lies between these extremes.

The Ionian intellectual movement does not stand alone. It is only one of several sporadic indications of the working of some powerful mental ferment over the whole of the area comprised between the Ægean and Northern Hindostan during the eighth, seventh, and sixth centuries before our era. In these three hundred years, prophetism attained its apogee among the Semites of Palestine; Zoroasterism grew and became the creed of a conquering race, the Iranic Aryans; Buddhism rose and spread with marvellous rapidity among the Aryans of Hindostan; while scientific naturalism took its rise among the Aryans of Ionia. It would be difficult to find another three centuries which have given birth to four events of equal importance. All the principal existing religions of mankind have grown out of the first three : while the fourth is the little spring, now swollen into the great stream of positive science. So far as physical possibilities go, the prophet Jeremiah and the oldest Ionian philosopher might have met and conversed. If they had done so, they would probably have disagreed a good deal; and it is interesting to reflect that their discussions might have

embraced questions which, at the present day, are
still hotly controverted.

The old Ionian philosophy, then, seems to be only
one of many results of a stirring of the moral and
intellectual life of the Aryan and the Semitic popu-
lations of Western Asia. The conditions of this
general awakening were doubtless manifold ; but
there is one which modern research has brought into
great prominence. This is the existence of extremely
ancient and highly advanced societies in the valleys
of the Euphrates and of the Nile.

It is now known that, more than a thousand —
perhaps more than two thousand—years before the
sixth century B.C., civilization had attained a relatively high pitch among the Babylonians and the
Egyptians. Not only had painting, sculpture,
architecture, and the industrial arts reached a remarkable development ; but in Chaldæa, at any rate,
a vast amount of knowledge had been accumulated
and methodized, in the departments of grammar,
mathematics, astronomy, and natural history. Where
such traces of the scientific spirit are visible,
naturalistic speculation is rarely far off, though, so
far as I know, no remains of an Accadian, or
Egyptian, philosophy, properly so called, have yet
been recovered.

Geographically, Chaldæa occupied a central position among the oldest seats of civilization. Commerce, largely aided by the intervention of those
colossal pedlars, the Phœnicians, had brought Chaldæa
into connection with all of them, for a thousand
years before the epoch at present under consideration.

And in the ninth, eighth and seventh centuries, the Assyrian, the depositary of Chaldæan civilization, as the Macedonian and the Roman, at a later date, were the depositaries of Greek culture, had added irresistible force to the other agencies for the wide distribution of Chaldæan literature, art, and science.

I confess that I find it difficult to imagine that the Greek immigrants—who stood in somewhat the same relation to the Babylonians and the Egyptians as the later Germanic barbarians to the Romans of the Empire—should not have been immensely influenced by the new life with which they became acquainted. But there is abundant direct evidence of the magnitude of this influence in certain spheres. I suppose it is not doubted that the Greek went to school with the Oriental for his primary instruction in reading, writing, and arithmetic ; and that Semitic theology supplied him with some of his mythological lore. Nor does there now seem to be any question about the large indebtedness of Greek art to that of Chaldæa and that of Egypt.

But the manner of that indebtedness is very instructive. The obligation is clear, but its limits are no less definite. Nothing better exemplifies the indomitable originality of the Greeks than the relations of their art to that of the Orientals. Far from being subdued into mere imitators by the technical excellence of their teachers, they lost no time in bettering the instruction they received, using their models as mere stepping stones on the way to those unsurpassed and unsurpassable achievements which are all their own. The shibboleth of Art is the human figure. The ancient Chaldæans and Egyptians, like the modern Japanese, did wonders in the representation of birds and quadrupeds; they even attained to something more than respectability in human portraiture. But their utmost efforts never brought them within range of the best Greek embodiments of the grace of womanhood, or of the severer beauty of manhood.

It is worth while to consider the probable effect upon the acute and critical Greek mind of the conflict of ideas, social, political, and theological, which arose out of the conditions of life in the Asiatic colonies. The Ionian polities had passed through the whole gamut of social and political changes, from patriarchal and occasionally oppressive kingship to rowdy and still more burdensome mobship—no doubt with infinitely eloquent and copious argumentation, on both sides, at every stage of their progress towards that arbitrament of force which settles most political questions. The marvellous speculative faculty, latent in the Ionian, had come in contact with Mesopotamian, Egyptian, Phœnician theologies and cosmogonies; with the illuminati of Orphism and the fanatics and dreamers of the Mysteries; possibly with Buddhism and Zoroasterism; possibly even with Judaism. And it has been observed that the mutual contradictions of antagonistic supernaturalisms are apt to play a large part among the generative agencies of naturalism.

Thus, various external influences may have contributed to the rise of philosophy among the Ionian Greeks of the sixth century. But the assimilative

capacity of the Greek mind—its power of Hellenizing
whatever it touched—has here worked so effectually,
that, so far as I can learn, no indubitable traces of
such extraneous contributions are now allowed to exist
by the most authoritative historians of Philosophy.
Nevertheless, I think it must be admitted that the
coincidences between the Heracleito-stoical doctrines
and those of the older Hindu philosophy are
extremely remarkable.   In both, the cosmos pursues
an eternal succession of cyclical changes.   The great
year, answering to the Kalpa, covers an entire cycle
from the origin of the universe as a fluid to its
dissolution in fire—"Humor initium, ignis exitus
mundi," as Seneca has it.   In both systems, there is
immanent in the cosmos a source of energy, Brahma,
or the Logos, which works according to fixed laws.
The individual soul is an efflux of this world-spirit,
and returns to it.   Perfection is attainable only by
individual effort, through ascetic discipline, and is
rather a state of painlessness than of happiness; if
indeed it can be said to be a state of anything, save
the negation of perturbing emotion.   The hatchment
motto "In Cœlo Quies" would serve both Hindu and
Stoic ; and absolute quiet is not easily distinguishable
from annihilation.

Zoroasterism, which, geographically, occupies a
position intermediate between Hellenism and
Hinduism, agrees with the latter in recognizing the
essential evil of the cosmos ; but differs from both in
its intensely anthropomorphic personification of the
two antagonistic principles, to the one of which it
ascribes all the good ; and, to the other, all the evil.

In fact, it assumes the existence of two worlds, one good and one bad ; the latter created by the evil power for the purpose of damaging the former. The existing cosmos is a mere mixture of the two, and the 'last judgment' is a root-and-branch extirpation of the work of Ahriman.

Note 12 (p. 69).

There is no snare in which the feet of a modern student of ancient lore are more easily entangled, than that which is spread by the similarity of the language of antiquity to modern modes of expression. I do not presume to interpret the obscurest of Greek philosophers ; all I wish is to point out, that his words, in the sense accepted by competent interpreters, fit modern ideas singularly well.

So far as the general theory of evolution goes there is no difficulty. The aphorism about the river ; the figure of the child playing on the shore ; the kingship and fatherhood of strife, seem decisive. The ὁδός ἄνω κάτω μίη expresses, with singular aptness, the cyclical aspect of the one process of organic evolution in individual plants and animals : yet it may be a question whether the Heracleitean strife included any distinct conception of the struggle for existence. Again, it is tempting to compare the part played by the Heracleitean 'fire' with that ascribed by the moderns to heat, or rather to that cause of motion of which heat is one expression ; and a little ingenuity might find a foreshadowing of the doctrine of the conservation of energy, in the saying that all the things are changed into fire and fire into all things, as gold into goods and goods into gold.

Note 13 (p. 71).

Popes lines in the *Essay on Man* (Ep. i. 267–8),

"All are but parts of one stupendous whole,
Whose body Nature is, and God the soul,"

simply paraphrase Seneca's "quem in hoc mundo locum deus obtinet, hunc in homine animus: quod est illic materia, id nobis corpus est."—(Ep. lxv. 24); which again is a Latin version of the old Stoical doctrine, εἰς ἅπαν τοῦ κόσμου μέρος διήκει ὁ νοῦς, καθάπερ ἀφ' ἡμῶν ἡ ψυχή.

So far as the testimony for the universality of what ordinary people call 'evil' goes, there is nothing better than the writings of the Stoics themselves. They might serve as a storehouse for the epigrams of the ultra-pessimists. Heracleitus (*circa* 500 B.C.) says just as hard things about ordinary humanity as his disciples centuries later; and there really seems no need to seek for the causes of this dark view of life in the circumstances of the time of Alexander's successors or of the early Emperors of Rome. To the man with an ethical ideal, the world, including himself, will always seem full of evil.

Note 14 (p. 73).

I use the well-known phrase, but decline responsibility for the libel upon Epicurus, whose doctrines were far less compatible with existence in a stye than those of the Cynics. If it were steadily borne in mind that the conception of the 'flesh' as the source of evil, and the great saying 'Initium est salutis notitia peccati,' are the property of Epicurus, fewer illusions about Epicureanism would pass muster for accepted truth.

Note 15 (p. 75).

The Stoics said that man was a ζῷον λογικὸν πολιτικὸν φιλάλληλον, or a rational, a political, and an altruistic or philanthropic animal. In their view, his higher nature tended to develop in these three directions, as a plant tends to grow up into its typical form. Since, without the introduction of any consideration of pleasure or pain, whatever thwarted the realization of its type by the plant might be said to be bad, and whatever helped it good ; so virtue, in the Stoical sense, as the conduct which tended to the attainment of the rational, political, and philanthropic ideal, was good in itself, and irrespectively of its emotional concomitants.

Man is an "animal sociale communi bono genitum." The safety of society depends upon practical recognition of the fact. "Salva autem esse societas nisi custodia et amore partium non possit," says Seneca. (*De. Ira,* ii. 31.)

Note 16 (p. 75).

The importance of the physical doctrine of the Stoics lies in its clear recognition of the universality of the law of causation, with its corollary, the order of nature : the exact form of that order is an altogether secondary consideration.

Many ingenious persons now appear to consider that the incompatibility of pantheism, of materialism, and of any doubt about the immortality of the soul, with religion and morality, is to be held as an axiomatic truth. I confess that I have a certain difficulty in accepting this dogma. For the Stoics were notoriously materialists and pantheists of the most extreme character ; and while no strict Stoic believed in the eternal duration of the individual soul, some even denied its persistence after death. Yet it is equally certain that of all gentile philosophies, Stoicism exhibits the highest ethical development, is animated by the most religious spirit, and has exerted the profoundest influence upon the moral and religious development not merely of the best men among the Romans, but among the moderns down to our own day.

Seneca was claimed as a Christian and placed among the saints by the fathers of the early Christian Church ; and the genuineness of a correspondence between him and the apostle Paul has been hotly maintained in our own time, by orthodox writers. That the letters, as we possess them, are worthless forgeries is obvious ; and writers as wide apart as Baur and Lightfoot agree that the whole story is devoid of foundation.

The dissertation of the late Bishop of Durham (*Epistle to the Philippians*) is particularly worthy of study, apart from this question, on account of the evidence which it supplies of the numerous similarities of thought between Seneca and the writer of the Pauline epistles. When it is remembered that the writer of the Acts puts a quotation from Aratus, or Cleanthes, into the mouth of the apostle ; and that Tarsus was a great seat of philosophical and especially stoical learning (Chrysippus himself was a native of the adjacent town of Sôli), there is no difficulty in understanding the origin of these resemblances. See, on this subject, Sir Alexander Grant's dissertation in his edition of *The Ethics of Aristotle* (where there is an interesting reference to the stoical character of Bishop Butler's ethics), the concluding pages of Dr. Weygoldt's instructive little work *Die Philosophie der Stoa*, and Aubertin's *Sénèque et Saint Paul*.

It is surprising that a writer of Dr. Lightfoot's stamp should speak of Stoicism as a philosophy of ' despair.' Surely, rather, it was a philosophy of men who, having cast off all illusions, and the childishness of despair among them, were minded to endure in patience whatever conditions the cosmic process might create, so long as those conditions were compatible with the progress towards virtue, which alone, for them, conferred a worthy object on existence. There is no note of despair in the stoical declaration that the perfected ' wise man ' is the equal of Zeus in everything but the duration of his existence. And, in my judgment, there is as little pride about it, often as it serves for the text of discourses on stoical arrogance. Grant the stoical postulate that there is no good except virtue ; grant that the perfected wise man is altogether virtuous, in consequence of being guided in all things by the reason, which is an effluence of Zeus, and there seems no escape from the stoical conclusion.

Note 17 (p. 76).

Our "Apathy" carries such a different set of connotations from its Greek original that I have ventured on using the latter as a technical term.

Note 18 (p. 77).

Many of the stoical philosophers recommended their disciples to take an active share in public affairs; and in the Roman world, for several centuries, the best public men were strongly inclined to Stoicism. Nevertheless, the logical tendency of Stoicism seems to me to be fulfilled only in such men as Diogenes and Epictetus.

Note 19 (p. 80).

"Criticisms on the Origin of Species," 1864. *Collected Essays*, vol. ii. p. 91. [1894.]

Note 20 (p. 81).

Of course, strictly speaking, social life, and the ethical process in virtue of which it advances towards perfection, are part and parcel of the general process of evolution, just as the gregarious habit of innumerable plants and animals, which has been of immense advantage to them, is so. A hive of bees is an organic polity, a society in which the part played by each member is determined by organic necessities. Queens, workers, and drones are, so to speak, castes, divided from one another by marked physical barriers. Among birds and mammals, societies are formed, of which the bond in many cases seems to be purely psychological ; that is to say, it appears to depend upon the liking of the individuals for one another's company The tendency of individuals to over self-assertion is kept down by fighting. Even in these rudimentary forms of society, love and fear come into play, and enforce a greater or less renunciation of self-will. To this extent the general cosmic process begins to be checked by a rudimentary ethical process, which is, strictly speaking, part of the former, just as the ' governor ' in a steam-engine is part of the mechanism of the engine.

Note 21 (p. 82).

See " Government : Anarchy or Regimentation," *Collected Essays*, vol. i. pp. 413—418. It is this form of political philosophy to which I conceive the epithet of ' reasoned savagery ' to be strictly applicable. [1894.]

Note 22 (p. 83).

" L'homme n'est qu'un roseau, le plus faible de la nature, mais c'est un roseau pensant. Il ne faut

pas que l'univers entier s'arme pour l'écraser.   Une
vapeur, une goutte d'eau, suffit pour le tuer.   Mais
quand l'univers l'écraserait, l'homme serait encore
plus noble que ce qui le tue, parce qu'il sait qu'il
meurt ; et l'avantage que l'univers a sur lui, l'univers
n'en sait rien."—*Pensées de Pascal.*

Note 23 (p. 85).

The use of the word " Nature" here may be criti-
cised.   Yet the manifestation of the natural tendencies
of men is so profoundly modified by training that it
is hardly too strong.   Consider the suppression of
the sexual instinct between near relations.

Note 24 (p. 86).

A great proportion of poetry is addressed by the
young to the young ; only the great masters of the
art are capable of divining, or think it worth while
to enter into, the feelings of retrospective age.   The
two great poets whom we have so lately lost, Tennyson
and Browning, have done this, each in his own
inimitable way ; the one in the *Ulysses*, from which
I have borrowed ; the other in that wonderful
fragment ' Childe Roland to the dark Tower came.'

III

SCIENCE AND MORALS

[1886]

IN spite of long and, perhaps, not unjustifiable
hesitation, I begin to think that there must be
something in telepathy. For evidence, which I
may not disregard, is furnished by the last number
of the "Fortnightly Review" that among the
hitherto undiscovered endowments of the human
species, there may be a power even more wonderful than the mystic faculty by which the esoteric-
ally Buddhistic sage " upon the farthest mountain
in Cathay " reads the inmost thoughts of a dweller
within the homely circuit of the London postal
district. Great indeed is the insight of such a
seer; but how much greater is his who combines
the feat of reading, not merely the thoughts of
which the thinker is aware, but those of which
he knows nothing; who sees him unconsciously
drawing the conclusions which he repudiates, and

supporting the doctrines which he detests. To reflect upon the confusion which the working of such a power as this may introduce into one's ideas of personality and responsibility is perilous —madness lies that way. But truth is truth, and I am almost fain to believe in this magical visibility of the non-existent when the only alternative is the supposition that the writer of the article on " Materialism and Morality " in vol. xl. (1886) of the " Fortnightly Review," in spite of his manifest ability and honesty, has pledged himself, so far as I am concerned, to what, if I may trust my own knowledge of my own thoughts, must be called a multitude of errors of the first magnitude.

I so much admire Mr. Lilly's outspokenness, I am so completely satisfied with the uprightness of his intentions, that it is repugnant to me to quarrel with anything he may say; and I sympathise so warmly with his manly scorn of the vileness of much that passes under the name of literature in these times, that I would willingly be silent under his by no means unkindly exposition of his theory of my own tenets, if I thought that such personal abnegation would serve the interest of the cause we both have at heart. But I cannot think so. My creed may be an ill-favoured thing, but it is mine own, as Touchstone says of his lady-love ; and I have so high an opinion of the solid virtues of the object of my affections that I cannot calmly see her personated by a wench who is much uglier and has no virtue worth speaking of. I hope I should be ready to stand by a falling cause if I had ever adopted it; but suffering for a falling cause, which one has done one's best to bring to the ground, is a kind of martyrdom for which I have no taste. In my opinion, the philosophical theory which Mr. Lilly attributes to me—but which I have over and over again disclaimed—is untenable and destined to extinction; and I not unreasonably demur to being counted among its defenders.

After the manner of a mediæval disputant, Mr. Lilly posts up three theses, which, as he conceives, embody the chief heresies propagated by the late Professor Clifford, Mr. Herbert Spencer, and myself. He says that we agree " (1) in putting aside, as unverifiable, everything which the senses cannot verify; (2) everything beyond the bounds of physical science; (3) everything which cannot be brought into a laboratory and dealt with chemically " (p. 578).

My lamented young friend Clifford, sweetest of natures though keenest of disputants, is out of reach of our little controversies, but his works speak for him, and those who run may read a refutation of Mr. Lilly's assertions in them. Mr. Herbert Spencer, hitherto, has shown no lack either of ability or of inclination to speak for himself; and it would be a superfluity, not to say an impertinence, on my part, to take up the cudgels for him. But, for myself, if my knowledge of my own consciousness may be assumed to be adequate (and I make not the least pretension to acquaintance with what goes on in my "Unbewusstsein"), I may be permitted to observe that the first proposition appears to me to be not true; that the second is in the same case; and that, if there be gradations in untrueness, the third is so monstrously untrue that it hovers on the verge of absurdity, even if it does not actually flounder in that logical limbo. Thus, to all three theses, I reply in appropriate fashion, *Nego*—I say No; and I proceed to state the grounds of that negation, which the proprieties do not permit me to make quite so emphatic as I could desire.

Let me begin with the first assertion, that I " put aside, as unverifiable, everything which the senses cannot verify." Can such a statement as this be seriously made in respect of any human being? But I am not appointed apologist for mankind in general; and confining my observations to myself, I beg leave to point out that, at this present moment, I entertain an unshakable conviction that Mr. Lilly is the victim of a patent and enormous misunderstanding, and that I have not the slightest intention of putting that conviction aside because I cannot " verify" it either by touch, or taste, or smell, or hearing, or sight, which (in the absence of any trace of telepathic faculty) make up the totality of my senses.

Again, I may venture to admire the clear and

vigorous English in which Mr. Lilly embodies his
views; but the source of that admiration does not
lie in anything which my five senses enable me to
discover in the pages of his article, and of which
an orang-outang might be just as acutely sensible.
No, it lies in an appreciation of literary form and
logical structure by æsthetic and intellectual
faculties which are not senses, and which are not
unfrequently sadly wanting where the senses are
in full vigour. My poor relation may beat me in
the matter of sensation; but I am quite confident
that, when style and syllogisms are to be dealt
with, he is nowhere.

If there is anything in the world which I do
firmly believe in, it is the universal validity of the
law of causation; but that universality cannot be
proved by any amount of experience, let alone
that which comes to us through the senses. And
when an effort of volition changes the current of
my thoughts, or when an idea calls up another
associated idea, I have not the slightest doubt
that the process to which the first of the phenomena, in each case, is due stands in the relation
of cause to the second. Yet the attempt to verify
this belief by sensation would be sheer lunacy.
Now I am quite sure that Mr. Lilly does not
doubt my sanity; and the only alternative seems
to be the admission that his first proposition is
erroneous.

The second thesis charges me with putting

aside " as unverifiable " " everything beyond the
bounds of physical science." Again I say, No.
Nobody, I imagine, will credit me with a desire
to limit the empire of physical science, but I really
feel bound to confess that a great many very
familiar and, at the same time, extremely important phenomena lie quite beyond its legitimate
limits. I cannot conceive, for example, how the
phenomena of consciousness, as such and apart
from the physical process by which they are
called into existence, are to be brought within
the bounds of physical science. Take the simplest
possible example, the feeling of redness. Physical
science tells us that it commonly arises as a consequence of molecular changes propagated from
the eye to a certain part of the substance of the
brain, when vibrations of the luminiferous ether
of a certain character fall upon the retina. Let
us suppose the process of physical analysis pushed
so far that one could view the last link of this
chain of molecules, watch their movements as if
they were billiard balls, weigh them, measure
them, and know all that is physically knowable
about them. Well, even in that case, we should
be just as far from being able to include the
resulting phenomenon of consciousness, the feeling
of redness, within the bounds of physical science,
as we are at present. It would remain as unlike
the phenomena we know under the names of
matter and motion as it is now. If there is any

plain truth upon which I have made it my
business to insist over and over again it is this—
and whether it is a truth or not, my insistence
upon it leaves not a shadow of justification for
Mr. Lilly's assertion. But I ask in this case also, how is it conceivable
that any man, in possession of all his natural
faculties, should hold such an opinion? I do not
suppose that I am exceptionally endowed because
I have all my life enjoyed a keen perception of
the beauty offered us by nature and by art. Now
physical science may and probably will, some day,
enable our posterity to set forth the exact physical
concomitants and conditions of the strange rapture
of beauty. But if ever that day arrives, the
rapture will remain, just as it is now, outside and
beyond the physical world; and, even in the
mental world, something superadded to mere sensation. I do not wish to crow unduly over my
humble cousin the orang, but in the æsthetic
province, as in that of the intellect, I am afraid
he is nowhere. I doubt not he would detect a
fruit amidst a wilderness of leaves where I could
see nothing; but I am tolerably confident that he
has never been awestruck, as I have been, by the
dim religious gloom, as of a temple devoted to the
earthgods, of the tropical forests which he in-
habits. Yet I doubt not that our poor long-
armed and short-legged friend, as he sits medita-
tively munching his durian fruit, has something

behind that sad Socratic face of his which is
utterly "beyond the bounds of physical science."
Physical science may know all about his clutching
the fruit and munching it and digesting it, and
how the physical titillation of his palate is transmitted to some microscopic cells of the gray
matter of his brain. But the feelings of sweet-
ness and of satisfaction which, for a moment, hang
out their signal lights in his melancholy eyes, are
as utterly outside the bounds of physics as is the
" fine frenzy " of a human rhapsodist.

Does Mr. Lilly really believe that, putting me
aside, there is any man with the feeling of music
in him who disbelieves in the reality of the delight
which he derives from it, because that delight
lies outside the bounds of physical science, not
less than outside the region of the mere sense of
hearing ?   But, it may be, that he includes music,
painting, and sculpture under the head of physical
science, and in that case I can only regret I am
unable to follow him in his ennoblement of my
favourite pursuits.

The third thesis runs that I put aside " as unverifiable " " everything which cannot be brought
into a laboratory and dealt with chemically ";
and, once more, I say No. This wondrous
allegation is no novelty; it has not unfrequently
reached me from that region where gentle (or
ungentle) dulness so often holds unchecked
sway—the pulpit. But I marvel to find that a

writer of Mr. Lilly's intelligence and good faith is willing to father such a wastrel. If I am to deal with the thing seriously, I find myself met by one of the two horns of a dilemma. Either some meaning, as unknown to usage as to the dictionaries, attaches to "laboratory" and "chemical," or the proposition is (what am I to say in my sore need for a gentle and yet appropriate word?)—well—unhistorical.

Does Mr. Lilly suppose that I put aside "as unverifiable" all the truths of mathematics, of philology, of history? And if I do not, will he have the great goodness to say how the binomial theorem is to be dealt with "chemically," even in the best-appointed "laboratory"; or where the balances and crucibles are kept by which the various theories of the nature of the Basque language may be tested; or what reagents will extract the truth from any given History of Rome, and leave the errors behind as a residual calx?

I really cannot answer these questions, and unless Mr. Lilly can, I think he would do well hereafter to think more than twice before attributing such preposterous notions to his fellow-men, who, after all, as a learned counsel said, are vertebrated animals.

The whole thing perplexes me much; and I am sure there must be an explanation which will leave Mr. Lilly's reputation for common sense

and fair dealing untouched. Can it be—I put
this forward quite tentatively—that Mr. Lilly is
the victim of a confusion, common enough among
thoughtless people, and into which he has fallen
unawares ? Obviously, it is one thing to say
that the logical methods of physical science are of
universal applicability, and quite another to affirm
that all subjects of thought lie within the province of physical science. I have often declared
my conviction that there is only one method by
which intellectual truth can be reached, whether
the subject-matter of investigation belongs to the
world of physics or to the world of consciousness ;
and one of the arguments in favour of the use of
physical science as an instrument of education
which I have oftenest used is that, in my opinion,
it exercises young minds in the appreciation of
inductive evidence better than any other study.
But while I repeat my conviction that the physical
sciences probably furnish the best and most easily
appreciable illustrations of the one and indivisible
mode of ascertaining truth by the use of reason,
I beg leave to add that I have never thought of
suggesting that other branches of knowledge may
not afford the same discipline ; and assuredly I
have never given the slightest ground for the
attribution to me of the ridiculous contention
that there is nothing true outside the bounds of
physical science. Doubtless people who wanted
to say something damaging, without too nice a

regard to its truth or falsehood, have often
enough misrepresented my plain meaning. But
Mr. Lilly is not one of these folks at whom one
looks and passes by, and I can but sorrowfully
wonder at finding him in such company.

So much for the three theses which Mr. Lilly
has nailed on to the page of this Review. I think
I have shown that the first is inaccurate, that the
second is inaccurate, and that the third is inaccurate; and that these three inaccurates constitute one prodigious, though I doubt not unintentional, misrepresentation. If Mr. Lilly and I
were dialectic gladiators, fighting in the arena of
the "Fortnightly," under the eye of an editorial
lanista, for the delectation of the public, my best
tactics would now be to leave the field of battle.
For the question whether I do, or do not, hold
certain opinions is a matter of fact, with regard to
which my evidence is likely to be regarded as
conclusive—at least until such time as the telepathy of the unconscious is more generally recognised.

However, some other assertions are made by
Mr. Lilly which more or less involve matters of
opinion whereof the rights and wrongs are less
easily settled, but in respect of which he seems to
me to err quite as seriously as about the topics
we have been hitherto discussing. And the importance of these subjects leads me to venture upon
saying something about them, even though I am

thereby compelled to leave the safe ground of personal knowledge.

Before launching the three torpedoes which have so sadly exploded on board his own ship, Mr. Lilly says that with whatever " rhetorical ornaments I may gild my teaching," it is " Materialism." Let me observe, in passing, that rhetorical ornament is not in my way, and that gilding refined gold would, to my mind, be less objectionable than varnishing the fair face of truth with that pestilent cosmetic, rhetoric. If I believed that I had any claim to the title of " Materialist," as that term is understood in the language of philosophy and not in that of abuse, I should not attempt to hide it by any sort of gilding. I have not found reason to care much for hard names in the course of the last thirty years, and I am too old to develop a new sensitiveness. But, to repeat what I have more than once taken pains to say in the most unadorned of plain language, I repudiate, as philosophical error, the doctrine of Materialism as I understand it, just as I repudiate the doctrine of Spiritualism as Mr. Lilly presents it, and my reason for thus doing is, in both cases, the same; namely, that, whatever their differences, Materialists and Spiritualists agree in making very positive assertions about matters of which I am certain I know nothing, and about which I believe they are, in truth, just as ignorant. And further, that, even when their

assertions are confined to topics which lie within
the range of my faculties, they often appear to
me to be in the wrong. And there is yet another
reason for objecting to be identified with either of
these sects; and that is that each is extremely
fond of attributing to the other, by way of re-
proach, conclusions which are the property of
neither, though they infallibly flow from the
logical development of the first principles of both.
Surely a prudent man is not to be reproached
because he keeps clear of the squabbles of these
philosophical Bianchi and Neri, by refusing to
have anything to do with either?

I understand the main tenet of Materialism to
be that there is nothing in the universe but
matter and force; and that all the phenomena of
nature are explicable by deduction from the properties assignable to these two primitive factors.
That great champion of Materialism whom Mr.
Lilly appears to consider to be an authority in
physical science, Dr. Büchner, embodies this
article of faith on his title-page. *Kraft und Stoff*
—force and matter—are paraded as the Alpha and
Omega of existence. This I apprehend is the
fundamental article of the faith materialistic;
and whosoever does not hold it is condemned by
the more zealous of the persuasion (as I have
some reason to know) to the Inferno appointed
for fools or hypocrites. But all this I heartily
disbelieve; and at the risk of being charged with

wearisome repetition of an old story, I will briefly give my reasons for persisting in my infidelity. In the first place, as I have already hinted, it seems to me pretty plain that there is a third thing in the universe, to wit, consciousness, which, in the hardness of my heart or head, I cannot see to be matter, or force, or any conceivable modification of either, however intimately the manifestations of the phenomena of consciousness may be connected with the phenomena known as matter and force. In the second place, the arguments used by Descartes and Berkeley to show that our certain knowledge does not extend beyond our states of consciousness, appear to me to be as irrefragable now as they did when I first became acquainted with them some half-century ago. All the materialistic writers I know of who have tried to bite that file have simply broken their teeth. But, if this is true, our one certainty is the existence of the mental world, and that of *Kraft und Stoff* falls into the rank of, at best, a highly probable hypothesis.

Thirdly, when I was a mere boy, with a perverse tendency to think when I ought to have been playing, my mind was greatly exercised by this formidable problem, What would become of things if they lost their qualities? As the qualities had no objective existence, and the thing without qualities was nothing, the solid world seemed whittled away—to my great horror. As I grew

older, and learned to use the terms matter and
force, the boyish problem was revived, *mutato
nomine*. On the one hand, the notion of matter
without force seemed to resolve the world into a
set of geometrical ghosts, too dead even to jabber.
On the other hand, Boscovich's hypothesis, by
which matter was resolved into centres of force,
was very attractive. But when one tried to think
it out, what in the world became of force considered as an objective entity? Force, even the
most materialistic of philosophers will agree with
the most idealistic, is nothing but a name for the
cause of motion. And if, with Boscovich, I
resolved things into centres of force, then matter
vanished altogether and left immaterial entities
in its place. One might as well frankly accept
Idealism and have done with it.

I must make a confession, even if it be humili-
ating. I have never been able to form the
slightest conception of those " forces " which the
Materialists talk about, as if they had samples of
them many years in bottle. They tell me that
matter consists of atoms, which are separated by
mere space devoid of contents; and that, through
this void, radiate the attractive and repulsive
forces whereby the atoms affect one another. If
anybody can clearly conceive the nature of these
things which not only exist in nothingness, but
pull and push there with great vigour, I envy
him for the possession of an intellect of larger
grasp, not only than mine, but than that of

Leibnitz or of Newton.[1] To me the " chimæra, bombinans in vacuo quia comedit secundas intentiones" of the schoolmen is a familiar and domestic creature compared with such " forces." Besides, by the hypothesis, the forces are not matter; and thus all that is of any particular consequence in the world turns out to be not matter on the Materialist's own showing. Let it not be supposed that I am casting a doubt upon the propriety of the employment of the terms " atom " and " force," as they stand among the working hypotheses of physical science. As formulæ which can be applied, with perfect precision and great convenience, in the interpretation of nature, their value is incalculable; but, as real entities, having an objective existence, an indivisible particle which nevertheless occupies space is surely inconceivable; and with respect to the operation of that atom, where it is not, by the aid of a " force" resident in nothingness, I am as little able to imagine it as I fancy any one else is.

Unless and until anybody will resolve all these doubts and difficulties for me, I think I have a right to hold aloof from Materialism. As to Spiritualism, it lands me in even greater difficul-

---

[1] See the famous *Collection of Papers*, published by Clarke in 1717. Leibnitz says: " 'Tis also a supernatural thing that bodies should *attract* one another at a distance without any intermediate means." And Clarke, on behalf of Newton, caps this as follows: " That one body should attract another without any intermediate *means* is, indeed, not a *miracle*, but a contradiction; for 'tis supposing something to act where it is not."

ties when I want to get change for its notes-of-hand in the solid coin of reality. For the assumed substantial entity, spirit, which is supposed to underlie the phenomena of consciousness, as matter underlies those of physical nature, leaves not even a geometrical ghost when these phenomena are abstracted. And, even if we suppose the existence of such an entity apart from qualities—that is to say, a bare existence—for mind, how does anybody know that it differs from that other entity, apart from qualities, which is the supposed substratum of matter? Spiritualism is, after all, little better than Materialism turned upside down. And if I try to think of the " spirit " which a man, by this hypothesis, carries about under his hat, as something devoid of relation to space, and as something indivisible, even in thought, while it is, at the same time, supposed to be in that place and to be possessed of half a dozen different faculties, I confess I get quite lost.

As I have said elsewhere, if I were forced to choose between Materialism and Idealism, I should elect for the latter; and I certainly would have nothing to do with the effete mythology of Spiritualism. But I am not aware that I am under any compulsion to choose either the one or the other. I have always entertained a strong suspicion that the sage who maintained that man is the measure of the universe was sadly in the wrong; and age and experience have not weakened

that conviction. In following these lines of speculation I am reminded of the quarter-deck walks of
my youth. In taking that form of exercise you
may perambulate through all points of the com-
pass with perfect safety, so long as you keep within
certain limits : forget those limits, in your ardour,
and mere smothering and spluttering, if not worse,
await you. I stick by the deck and throw a life-
buoy now and then to the struggling folk who
have gone overboard; and all I get for my
humanity is the abuse of all whenever they leave
off abusing one another.

Tolerably early in life I discovered that one of
the unpardonable sins, in the eyes of most people,
is for a man to presume to go about unlabelled.
The world regards such a person as the police do
an unmuzzled dog, not under proper control. I
could find no label that would suit me, so, in my
desire to range myself and be respectable, I invented one; and, as the chief thing I was sure of
was that I did not know a great many things that
the —ists and the —ites about me professed to be
familiar with, I called myself an Agnostic. Surely
no denomination could be more modest or more
appropriate; and I cannot imagine why I should
be every now and then haled out of my refuge
and declared sometimes to be a Materialist, sometimes an Atheist, sometimes a Positivist; and
sometimes, alas and alack, a cowardly or reaction-
ary Obscurantist.

I trust that I have, at last, made my case clear, and that henceforth I shall be allowed to rest in peace—at least, after a further explanation or two, which Mr. Lilly proves to me may be necessary. It has been seen that my excellent critic has original ideas respecting the meaning of the words "laboratory" and "chemical"; and, as it appears to me, his definition of "Materialist" is quite as much peculiar to himself. For, unless I misunderstand him, and I have taken pains not to do so, he puts me down as a Materialist (over and above the grounds which I have shown to have no foundation); firstly, because I have said that consciousness is a function of the brain; and, secondly, because I hold by determinism. With respect to the first point, I am not aware that there is any one who doubts that, in the proper physiological sense of the word function, consciousness, in certain forms at any rate, is a cerebral function. In physiology we call function that effect, or series of effects, which results from the activity of an organ. Thus, it is the function of muscle to give rise to motion; and the muscle gives rise to motion when the nerve which supplies it is stimulated. If one of the nerve-bundles in a man's arm is laid bare and a stimulus is applied to certain of the nervous filaments, the result will be production of motion in that arm. If others are stimulated, the result will be the production of the state of consciousness called pain. Now, if I trace these last nerve-filaments, I find them to be ultimately connected with part of the substance of the brain, just as the others turn out to be connected with muscular substance. If the production of motion in the one case is properly said to be the function of the muscular substance, why is the production of a state of consciousness in the other case not to be called a function of the cerebral substance? Once upon a time, it is true, it was supposed that a certain "animal spirit" resided in muscle and was the real active agent. But we have done with that wholly superfluous fiction so far as the muscular organs are concerned. Why are we to retain a corresponding fiction for the nervous organs?

If it is replied that no physiologist, however spiritual his leanings, dreams of supposing that simple sensations require a "spirit" for their production, then I must point out that we are all agreed that consciousness is a function of matter, and that particular tenet must be given up as a mark of Materialism. Any further argument will turn upon the question, not whether consciousness is a function of the brain, but whether all forms of consciousness are so. Again, I hold it would be quite correct to say that material changes are the causes of psychical phenomena (and, as a consequence, that the organs in which these changes take place have the production of such phenomena for their function), even if the spiritualistic hypothesis had any foundation. For nobody hesitates to say that an event A is the cause of an event Z, even if there are as many intermediate terms, known and unknown, in the chain of causation as there are letters between A and Z. The man who pulls the trigger of a loaded pistol placed close to another's head certainly is the cause of that other's death, though, in strictness, he "causes" nothing but the movement of the finger upon the trigger. And, in like manner, the molecular change which is brought about in a certain portion of the cerebral substance by the stimulation of a remote part of the body would be properly said to be the cause of the consequent feeling, whatever unknown terms were interposed between the physical agent and the actual psychical product. Therefore, unless Materialism has the monopoly of the right use of language, I see nothing materialistic in the phraseology which I have employed.

The only remaining justification which Mr. Lilly offers for dubbing me a Materialist, *malgré moi*, arises out of a passage which he quotes, in which I say that the progress of science means the extension of the province of what we call matter and force, and the concomitant gradual banishment from all regions of human thought of what we call spirit and spontaneity. I hold that opinion now,

if anything, more firmly than I did when I gave
utterance to it a score of years ago, for it has
been justified by subsequent events.   But what
that opinion has to do with Materialism I fail to
discover.   In my judgment, it is consistent with
the most thorough-going Idealism, and the
grounds of that judgment are really very plain
and simple.

The growth of science, not merely of physical
science, but of all science, means the demonstration
of order and natural causation among phenomena
which had not previously been brought under those
conceptions.   Nobody who is acquainted with the
progress of scientific thinking in every department
of human knowledge, in the course of the last two
centuries, will be disposed to deny that immense
provinces have been added to the realm of science ;
or to doubt that the next two centuries will be
witnesses of a vastly greater annexation.   More
particularly in the region of the physiology of the
nervous system is it justifiable to conclude from
the progress that has been made in analysing the
relations between material and psychical phenomena, that vast further advances will be made ;
and that, sooner or later, all the so-called spon-
taneous operations of the mind will have, not only
their relations to one another, but their relations
to physical phenomena, connected in natural series
of causes and effects, strictly defined.   In other
words, while, at present, we know only the nearer

moiety of the chain of causes and effects, by which the phenomena we call material give rise to those which we call mental; hereafter, we shall get to the further end of the series.

In my innocence, I have been in the habit of supposing that this is merely a statement of facts, and that the good Bishop Berkeley, if he were alive, would find such facts fit into his system without the least difficulty. That Mr. Lilly should play into the hands of his foes, by declaring that unmistakable facts make for them, is an exemplification of ways that are dark, quite unintelligible to me. Surely Mr. Lilly does not hold that the disbelief in spontaneity—which term, if it has any meaning at all, means uncaused action —is a mark of the beast Materialism? If so, he must be prepared to tackle many of the Cartesians (if not Descartes himself), Spinoza and Leibnitz among the philosophers, Augustine, Thomas Aquinas, Calvin and his followers among theologians, as Materialists—and that surely is a sufficient *reductio ad absurdum* of such a classification.

The truth is, that in his zeal to paint " Materialism," in large letters, on everything he dislikes, Mr. Lilly forgets a very important fact, which, however, must be patent to every one who has paid attention to the history of human thought; and that fact is, that every one of the speculative difficulties which beset Kant's three problems, the existence of a Deity, the freedom of the will, and immortality, existed ages before anything that can be called physical science, and would continue to exist if modern physical science were swept away. All that physical science has done has been to make, as it were, visible and tangible some difficulties that formerly were more hard of apprehension. Moreover, these difficulties exist just as much on the hypothesis of Idealism as on that of Materialism.

The student of nature, who starts from the axiom of the universality of the law of causation, cannot refuse to admit an eternal existence; if he admits the conservation of energy, he cannot deny the possibility of an eternal energy; if he admits the existence of immaterial phenomena in the form of consciousness, he must admit the possibility, at any rate, of an eternal series of such phenomena; and, if his studies have not been barren of the best fruit of the investigation of nature, he will have enough sense to see that when Spinoza says, "Per Deum intelligo ens absolute infinitum, hoc est substantiam constantem infinitis attributis," the God so conceived is one that only a very great fool would deny, even in his heart. Physical science is as little Atheistic as it is Materialistic.

So with respect to immortality. As physical science states this problem, it seems to stand thus: "Is there any means of knowing whether the series of states of consciousness, which has been

casually associated for threescore years and ten
with the arrangement and movements of innumerable millions of successively different material molecules, can be continued, in like association, with some substance which has not . the
properties of matter and force ? " As Kant said,
on a like occasion, if anybody can answer that
question, he is just the man I want to see. If he
says that consciousness cannot exist, except in
relation of cause and effect with certain organic
molecules, I must ask how he knows that; and if
he says it can, I must put the same question.
And I am afraid that, like jesting Pilate, I shall
not think it worth while (having but little time
before me) to wait for an answer.

Lastly, with respect to the old riddle of the
freedom of the will. In the only sense in which
the word freedom is intelligible to me—that is to
say, the absence of any restraint upon doing what
one likes within certain limits—physical science
certainly gives no more ground for doubting it
than the common sense of mankind does. And if
physical science, in strengthening our belief in the
universality of causation and abolishing chance as
an absurdity, leads to the conclusions of determinism, it does no more than follow the track of
consistent and logical thinkers in philosophy and
in theology, before it existed or was thought of.
Whoever accepts the universality of the law of
causation as a dogma of philosophy, denies the

existence of uncaused phenomena. And the
essence of that which is improperly called the
freewill doctrine is that occasionally, at any rate,
human volition is self-caused, that is to say, not
caused at all; for to cause oneself one must have
anteceded oneself—which is, to say the least of it,
difficult to imagine.

Whoever accepts the existence of an omniscient
Deity as a dogma of theology, affirms that the
order of things is fixed from eternity to eternity;
for the fore-knowledge of an occurrence means
that the occurrence will certainly happen; and
the certainty of an event happening is what is
meant by its being fixed or fated.[1]

[1] I may cite, in support of this obvious conclusion of sound
reasoning, two authorities who will certainly not be regarded
lightly by Mr. Lilly. These are Augustine and Thomas
Aquinas. The former declares that "Fate" is only an ill-
chosen name for Providence.

"Prorsus divina providentia regna constituuntur humana.
Quæ si propterea quisquam fato tribuit, quia ipsam Dei volun-
tatem vel potestatem fati nomine appellat, *sententiam teneat,
linguam corrigat*" (Augustinus *De Civitate Dei*, V. c. i.)

The other great doctor of the Catholic Church, "Divus
Thomas," as Suarez calls him, whose marvellous grasp and
subtlety of intellect seem to me to be almost without a parallel,
puts the whole case into a nutshell, when he says that the
ground for doing a thing in the mind of the doer is as it were
the pre-existence of the thing done :

"Ratio autem alicujus fiendi in mente actoris existens est
quædam præ-existentia rei fiendæ in eo" (*Summa*, Qu. xxiii.
Art. i.)

If this is not enough, I may further ask what "Materialist"
has ever given a better statement of the case for determinism,
on theistic grounds, than is to be found in the following passage
of the *Summa*, Qu. xiv. Art. xiii.

"Omnia quæ sunt in tempore, sunt Deo ab æterno præsentia,
non solum ea ex ratione quâ habet rationes rerum apud se

Whoever asserts the existence of an omnipotent Deity, that he made and sustains all things, and is the *causa causarum*, cannot, without a contradiction in terms, assert that there is any cause independent of him; and it is a mere subterfuge to assert that the cause of all things can " permit " one of these things to be an independent cause.

Whoever asserts the combination of omniscience and omnipotence as attributes of the Deity, does implicitly assert predestination. For he who knowingly makes a thing and places it in circumstances the operation of which on that thing he is perfectly acquainted with, does predestine that thing to whatever fate may befall it.

Thus, to come, at last, to the really important part of all this discussion, if the belief in a God is essential to morality, physical science offers no obstacle thereto; if the belief in immortality is essential to morality, physical science has no more to say against the probability of that doctrine than the most ordinary experience has, and it effectually closes the mouths of those who pretend to refute it by objections deduced from merely physical

presentes, ut quidam dicunt, sed quia ejus intuitus fertur ab æterno supra omnia, prout sunt in sua præsentialitate. *Unde manifestum est quod contingentia infallibiliter a Deo cognoscuntur*, in quantum subduntur divino conspectui secundum suam præsentialitatem; et tamen sunt futura contingentia, suis causis proximis comparata."

[As I have not said that Thomas Aquinas is professedly a determinist, I do not see the bearing of citations from him which may be more or less inconsistent with the foregoing.]

data. Finally, if the belief in the uncausedness of volition is essential to morality, the student of physical science has no more to say against that absurdity than the logical philosopher or theologian. Physical science, I repeat, did not invent determinism, and the deterministic doctrine would stand on just as firm a foundation as it does if there were no physical science. Let any one who doubts this read Jonathan Edwards, whose demonstrations are derived wholly from philosophy and theology.

Thus, when Mr. Lilly, like another Solomon Eagle, goes about proclaiming "Woe to this wicked city," and denouncing physical science as the evil genius of modern days—mother of materialism, and fatalism, and all sorts of other condemnable isms—I venture to beg him to lay the blame on the right shoulders; or, at least, to put in the dock, along with Science, those sinful sisters of hers, Philosophy and Theology, who, being so much older, should have known better than the poor Cinderella of the schools and universities over which they have so long dominated. No doubt modern society is diseased enough; but then it does not differ from older civilisations in that respect. Societies of men are fermenting masses, and, as beer has what the Germans call "Oberhefe" and "Unterhefe," so every society that has existed has had its scum at the top and its dregs at the bottom; but I doubt if any of the

" ages of faith " had less scum or less dregs, or even showed a proportionally greater quantity of sound wholesome stuff in the vat. I think it would puzzle Mr. Lilly, or any one else, to adduce convincing evidence that, at any period of the world's history, there was a more widespread sense of social duty, or a greater sense of justice, or of the obligation of mutual help, than in this England of ours. Ah! but, says Mr. Lilly, these are all products of our Christian inheritance ; when Christian dogmas vanish virtue will disappear too, and the ancestral ape and tiger will have full play. But there are a good many people who think it obvious that Christianity also inherited a good deal from Paganism and from Judaism; and that, if the Stoics and the Jews revoked their bequest, the moral property of Christianity would realise very little. And, if morality has survived the stripping off of several sets of clothes which have been found to fit badly, why should it not be able to get on very well in the light and handy garments which Science is ready to provide ?

But this by the way. If the diseases of society consist in the weakness of its faith in the existence of the God of the theologians, in a future state, and in uncaused volitions, the indication, as the doctors say, is to suppress Theology and Philosophy, whose bickerings about things of which they know nothing have been the prime cause and continual sustenance of that evil scepticism which is the Nemesis of meddling with the unknowable.

Cinderella is modestly conscious of her ignorance of these high matters. She lights the fire, sweeps the house, and provides the dinner; and is rewarded by being told that she is a base creature, devoted to low and material interests. But in her garret she has fairy visions out of the ken of the pair of shrews who are quarrelling down stairs. She sees the order which pervades the seeming disorder of the world; the great drama of evolution, with its full share of pity and terror, but also with abundant goodness and beauty, unrolls itself before her eyes; and she learns, in her heart of hearts, the lesson, that the foundation of morality is to have done, once and for all, with lying; to give up pretending to believe that for which there is no evidence, and repeating unintelligible propositions about things beyond the possibilities of knowledge.

She knows that the safety of morality lies neither in the adoption of this or that philosophical speculation, or this or that theological creed, but in a real and living belief in that fixed order of nature which sends social disorganisation upon the track of immorality, as surely as it sends physical disease after physical trespasses. And of that firm and lively faith it is her high mission to be the priestess.

GREAT BOOKS IN PHILOSOPHY PAPERBACK SERIES

ESTHETICS

- ❑ Aristotle—*The Poetics*
- ❑ Aristotle—*Treatise on Rhetoric*

ETHICS

- ❑ Aristotle—*The Nicomachean Ethics*
- ❑ Marcus Aurelius—*Meditations*
- ❑ Jeremy Bentham—*The Principles of Morals and Legislation*
- ❑ John Dewey—*Human Nature and Conduct*
- ❑ John Dewey—*The Moral Writings of John Dewey, Revised Edition*
- ❑ Epictetus—*Enchiridion*
- ❑ David Hume—*An Enquiry Concerning the Principles of Morals*
- ❑ Immanuel Kant—*Fundamental Principles of the Metaphysic of Morals*
- ❑ John Stuart Mill—*Utilitarianism*
- ❑ George Edward Moore—*Principia Ethica*
- ❑ Friedrich Nietzsche—*Beyond Good and Evil*
- ❑ Plato—*Protagoras, Philebus,* and *Gorgias*
- ❑ Bertrand Russell—*Bertrand Russell On Ethics, Sex, and Marriage*
- ❑ Arthur Schopenhauer—*The Wisdom of Life* and *Counsels and Maxims*
- ❑ Adam Smith—*The Theory of Moral Sentiments*
- ❑ Benedict de Spinoza—*Ethics* and *The Improvement of the Understanding*

LOGIC

- ❑ George Boole—*The Laws of Thought*

METAPHYSICS/EPISTEMOLOGY

- ❑ Aristotle—*De Anima*
- ❑ Aristotle—*The Metaphysics*
- ❑ Francis Bacon—*Essays*
- ❑ George Berkeley—*Three Dialogues Between Hylas and Philonous*
- ❑ W. K. Clifford—*The Ethics of Belief and Other Essays*
- ❑ René Descartes—*Discourse on Method* and *The Meditations*
- ❑ John Dewey—*How We Think*
- ❑ John Dewey—*The Influence of Darwin on Philosophy and Other Essays*
- ❑ Epicurus—*The Essential Epicurus: Letters, Principal Doctrines, Vatican Sayings, and Fragments*
- ❑ Sidney Hook—*The Quest for Being*
- ❑ David Hume—*An Enquiry Concerning Human Understanding*
- ❑ David Hume—*Treatise of Human Nature*
- ❑ William James—*The Meaning of Truth*
- ❑ William James—*Pragmatism*
- ❑ Immanuel Kant—*The Critique of Judgment*
- ❑ Immanuel Kant—*Critique of Practical Reason*
- ❑ Immanuel Kant—*Critique of Pure Reason*
- ❑ Gottfried Wilhelm Leibniz—*Discourse on Metaphysics* and the *Monadology*
- ❑ John Locke—*An Essay Concerning Human Understanding*
- ❑ George Herbert Mead—*The Philosophy of the Present*

- ❏ Charles S. Peirce—*The Essential Writings*
- ❏ Plato—*The Euthyphro, Apology, Crito,* and *Phaedo*
- ❏ Plato—*Lysis, Phaedrus,* and *Symposium*
- ❏ Bertrand Russell—*The Problems of Philosophy*
- ❏ George Santayana—*The Life of Reason*
- ❏ Sextus Empiricus—*Outlines of Pyrrhonism*
- ❏ Ludwig Wittgenstein—*Wittgenstein's Lectures: Cambridge, 1932–1935*

PHILOSOPHY OF RELIGION

- ❏ Jeremy Bentham—*The Influence of Natural Religion on the Temporal Happiness of Mankind*
- ❏ Marcus Tullius Cicero—*The Nature of the Gods* and *On Divination*
- ❏ Ludwig Feuerbach—*The Essence of Christianity*
- ❏ Paul Henry Thiry, Baron d'Holbach—*Good Sense*
- ❏ David Hume—*Dialogues Concerning Natural Religion*
- ❏ William James—*The Varieties of Religious Experience*
- ❏ John Locke—*A Letter Concerning Toleration*
- ❏ Lucretius—*On the Nature of Things*
- ❏ John Stuart Mill—*Three Essays on Religion*
- ❏ Friedrich Nietzsche—*The Antichrist*
- ❏ Thomas Paine—*The Age of Reason*
- ❏ Bertrand Russell—*Bertrand Russell On God and Religion*

SOCIAL AND POLITICAL PHILOSOPHY

- ❏ Aristotle—*The Politics*
- ❏ Mikhail Bakunin—*The Basic Bakunin: Writings, 1869–1871*
- ❏ Edmund Burke—*Reflections on the Revolution in France*
- ❏ John Dewey—*Freedom and Culture*
- ❏ John Dewey—*Individualism Old and New*
- ❏ John Dewey—*Liberalism and Social Action*
- ❏ G. W. F. Hegel—*The Philosophy of History*
- ❏ G. W. F. Hegel—*Philosophy of Right*
- ❏ Thomas Hobbes—*The Leviathan*
- ❏ Sidney Hook—*Paradoxes of Freedom*
- ❏ Sidney Hook—*Reason, Social Myths, and Democracy*
- ❏ John Locke—*Second Treatise on Civil Government*
- ❏ Niccolo Machiavelli—*The Prince*
- ❏ Karl Marx (with Friedrich Engels)—*The German Ideology,* including *Theses on Feuerbach and Introduction to the Critique of Political Economy*
- ❏ Karl Marx—*The Poverty of Philosophy*
- ❏ Karl Marx/Friedrich Engels—*The Economic and Philosophic Manuscripts of 1844* and *The Communist Manifesto*
- ❏ John Stuart Mill—*Considerations on Representative Government*
- ❏ John Stuart Mill—*On Liberty*
- ❏ John Stuart Mill—*On Socialism*
- ❏ John Stuart Mill—*The Subjection of Women*
- ❏ Montesquieu, Charles de Secondat—*The Spirit of Laws*

- ❏ Friedrich Nietzsche—*Thus Spake Zarathustra*
- ❏ Thomas Paine—*Common Sense*
- ❏ Thomas Paine—*Rights of Man*
- ❏ Plato—*Laws*
- ❏ Plato—*The Republic*
- ❏ Jean-Jacques Rousseau—*Émile*
- ❏ Jean-Jacques Rousseau—*The Social Contract*
- ❏ Mary Wollstonecraft—*A Vindication of the Rights of Men*
- ❏ Mary Wollstonecraft—*A Vindication of the Rights of Women*

GREAT MINDS PAPERBACK SERIES

ART

- ❏ Leonardo da Vinci—*A Treatise on Painting*

CRITICAL ESSAYS

- ❏ Desiderius Erasmus—*The Praise of Folly*
- ❏ Jonathan Swift—*A Modest Proposal and Other Satires*
- ❏ H. G. Wells—*The Conquest of Time*

ECONOMICS

- ❏ Charlotte Perkins Gilman—*Women and Economics:
  A Study of the Economic Relation between Women and Men*
- ❏ John Maynard Keynes—*The General Theory of Employment,
  Interest, and Money*
- ❏ John Maynard Keynes—*A Tract on Monetary Reform*
- ❏ Thomas R. Malthus—*An Essay on the Principle of Population*
- ❏ Alfred Marshall—*Money, Credit, and Commerce*
- ❏ Alfred Marshall—*Principles of Economics*
- ❏ Karl Marx—*Theories of Surplus Value*
- ❏ John Stuart Mill—*Principles of Political Economy*
- ❏ David Ricardo—*Principles of Political Economy and Taxation*
- ❏ Adam Smith—*Wealth of Nations*
- ❏ Thorstein Veblen—*Theory of the Leisure Class*

HISTORY

- ❏ Edward Gibbon—*On Christianity*
- ❏ Alexander Hamilton, John Jay, and James Madison—*The Federalist*
- ❏ Herodotus—*The History*
- ❏ Thucydides—*History of the Peloponnesian War*
- ❏ Andrew D. White—*A History of the Warfare of Science
  with Theology in Christendom*

LAW

- ❏ John Austin—*The Province of Jurisprudence Determined*

PSYCHOLOGY

❑ Sigmund Freud—*Totem and Taboo*

RELIGION

❑ Thomas Henry Huxley—*Agnosticism and Christianity and Other Essays*
❑ Ernest Renan—*The Life of Jesus*
❑ Upton Sinclair—*The Profits of Religion*
❑ Elizabeth Cady Stanton—*The Woman's Bible*
❑ Voltaire—*A Treatise on Toleration and Other Essays*

SCIENCE

❑ Jacob Bronowski—*The Identity of Man*
❑ Nicolaus Copernicus—*On the Revolutions of Heavenly Spheres*
❑ Marie Curie—*Radioactive Substances*
❑ Charles Darwin—*The Autobiography of Charles Darwin*
❑ Charles Darwin—*The Descent of Man*
❑ Charles Darwin—*The Origin of Species*
❑ Charles Darwin—*The Voyage of the* Beagle
❑ René Descartes—*Treatise of Man*
❑ Albert Einstein—*Relativity*
❑ Michael Faraday—*The Forces of Matter*
❑ Galileo Galilei—*Dialogues Concerning Two New Sciences*
❑ Ernst Haeckel—*The Riddle of the Universe*
❑ William Harvey—*On the Motion of the Heart and Blood in Animals*
❑ Werner Heisenberg—*Physics and Philosophy:
   The Revolution in Modern Science*
❑ Julian Huxley—*Evolutionary Humanism*
❑ Thomas H. Huxley—*Evolution and Ethics* and *Science and Morals*
❑ Edward Jenner—*Vaccination against Smallpox*
❑ Johannes Kepler—*Epitome of Copernican Astronomy*
   and *Harmonies of the World*
❑ Charles Mackay—*Extraordinary Popular Delusions*
   and the Madness of Crowds*
❑ James Clerk Maxwell—*Matter and Motion*
❑ Isaac Newton—*Opticks, Or Treatise of the Reflections,
   Inflections, and Colours of Light*
❑ Isaac Newton—*The Principia*
❑ Louis Pasteur and Joseph Lister—*Germ Theory and Its Application
   to Medicine* and *On the Antiseptic Principle of the Practice of Surgery*
❑ William Thomson (Lord Kelvin) and Peter Guthrie Tait—
   The Elements of Natural Philosophy
❑ Alfred Russel Wallace—*Island Life*

SOCIOLOGY

❑ Emile Durkheim—*Ethics and the Sociology of Morals*